AQA GCSE Biology

Editor: Graham Hill

Toby Houghton,
Christine Woodward

Hodder Murray
A MEMBER OF THE HODDER HEADLINE GROUP

The Publishers would like to thank the following for permission to reproduce copyright material:

p. 9 Science Photo Library/Saturn Stills; **p. 10** Corbis/Sean Aidan/Eye Ubiquitous: **p. 11** *t* Science Photo Library/AJ Photo, *c* Still Pictures/Michael J. Balick, *b* Corbis/Eric and David Hosking; **p. 12** Rex Features/Mike Webster; **p. 14** Bob Battersby; **p. 18** Anthony Blake Photo Library/Maximilian Stock Ltd; **p. 22** *l* Anthony Blake Photo Library/Joy Skipper, *c* Anthony Blake Photo Library/Maximilian Stock Ltd, *r* Bob Battersby; **p. 23** Science Photo Library/John Radcliffe Hospital; **p. 24** *l* Rex Features/Henryk T. Kaiser, *c* Rex Features/John Powell, *r* Rex Features/Image Source; **p. 25** *l* Corbis/Norbert Schaefer, *c* Rex Features/Image Source, *r* Alamy/Alaska Stock LLC; **p. 30** Rex Features/TS/Keystone USA; **p. 32** *tl* Corbis/Mediscan, *tr* Corbis/Visuals Unlimited, *bl* Science Photo Library; **p. 35** Rex Features/Garo/Phanie; **p. 38** Science Photo Library/Nancy Sefton; **p. 39** *c* Rex Features/John W. Warden, *b* Photolibrary/OSF/David Cayless; **p. 40** *tl* Photolibrary/OSF/Stephen Shepherd, *cl* Science Photo Library/John Beatty, *cr* Bruce Coleman Ltd, *bl* Ardea/Bob Gibbons; **p. 41** *l* Rex Features/Reso, *r* Alamy/John Sylvester; **p. 42** Alamy/Brian Elliott; **p. 43** Photodisc; **p. 44** Still Pictures/Mark Edwards; **p. 45** *l* Photolibrary/OSF/Harold Taylor, *cl* Photolibrary/OSF/Paulo De Oliveira, *cr* Still Pictures/Luc Vausort. *r* Science Photo Library/David Aubrey, *bl* Alamy/Vorldwide Picture Library; **p. 50** *t* Getty Images/David Woodfall, *bl* Christine Woodward, *br* Christine Woodward; **p. 51** Getty Images/AFP; **p. 52** Getty Images/Art Wolfe; **p. 54** Christine Woodward; **p. 56** *t* Christine Woodward, *b* Alamy/Jim Wcst; **p. 60** *tl* Science Photo Library/CAMR/A.B.Dowsett, /*r* Corbis/Stuart Westmorland, *bl* Rex Features/Richard Austin, *br* Rex Features/Rex Interstock; **p. 61** Science Photo Library/James King Holmes; **p. 64** *t* Science Photo Library/Kenneth W. Fink, *b* Science Photo Library/SCIMAT; **p. 65** Corbis/lan Harwood/Ecoscene; **p. 67** Science Photo Library/Michael Donne; **p. 68** Still Picturcs/Philippe Hays; **p. 69** *tl* Still Pictures/Martin Harvey, *tr* Science Photo Library/Gregory Dimijian, *c* Photolibrary/OSF/Sue Scott, *br* Mary Evans Picture Library: **p. 70** Corbis/Leonard de Selva; **p. 71** Hulton Archive/Getty Images; **p. 72** *t* Corbis/Reuters, *c* Corbis/Reuters; **p. 73** Science Photo Library/George Bernard; **p. 76** *tl* Science Photo Library/Andrew Syred, *tc* Science Photo Library/Susumu Nishinaga, *tr* Science Photo Library/Dr Torsten Wittmann, *bl* Science Photo Library/Manfred Kage, Peter Arnold INC., *br* Science Photo Library/Steve Gschmeissner; **p. 77** *l* Science Photo Library/Adam Hart-Davis, *c* Science Photo Library/Omikron, *b* Science Photo Library/Dr Gopal Murti; **p. 78** *t* Science Photo Library/Dr Jeremy Burgess, *b* Science Photo Library/Steve Gschmeissner; **p. 88** *main photo* Corbis/Stefan Puchner/dpa, *inset* Science Photo Library/Eye of Science; **p. 89** Science Photo Library/Biophoto Associates; **p. 95** Alamy/Holt Studios; **p. 97** *l* Alamy/Tom Mareachal, *tr* Corbis/Paul A Souders, *br* Corbis/Paul A. Souders; **p. 98** *l* Alamy/Phil Degginger, *b* Courtesy Wadsworth Control Systems; **p. 109** *main photo* Science Photo Library/CC Studio, *t inset* Grace/zefa/Corbis, *b inset* Science Photo Library/Prof. K. Seddon & Dr T. Evans, Queen's University, Belfast; **p. 113** Getty/Hulton Archive; **p. 119** Alamy/Phototake, Inc.; **p. 120** *bl* Anthony Blake Photo Library/Tony Robins, *br* Getty/Gary Buss; **p. 121** Lorna Ainger; **p. 122** Lorna Ainger; **p. 126** PYMCA/Manuela Zanotti; **p. 130** Corbis/Bettmann; **p. 133** Science Photo Library/L. Willatt, East Anglian Regional Genetics Service; **p. 137** Science Photo Library/Professor Miodrag Stojkovic; **p. 138** *tl* Science Photo Library/Professor Miodrag Stojkovic, *tc* Science Photo Library/Steve Gschmeissner, *tr* Science Photo Library/Susumu Nishinaga, *cl* Science Photo Library/Innerspace Imaging, *cr* Science Photo Library/Steve Gschmeissner, *bl* Science Photo Library/Steve Gschmeissner, *br* Science Photo Library/David McCarthy; **p. 140** PureStockX; **p. 141** Science Photo Library/CNRI; **p. 147** *tl* Science Photo Library/Mark Clarke, *tr* Corbis/Patrick Johns, *bl* Corbis/Walter Lockwood, *br* Science Photo Library/BSIP Laurent/H. Americain; **p. 152** Science Photo Library/Eye Of Science; **p. 159** Science Photo Library/Publiphoto Diffusion/Y. Beaulieu; **p. 162** Science Photo Library/Eye Of Science; **p. 163** *tl* Action Plus/ Neil Tingle, *tr* PurestockX; *bl* Alamy/Joanna Totolici, *br* Science Photo Library/Lea Paterson; **p. 168** Science Photo Library/Jerry Mason; **p. 169** Science Photo Library/Susumu Nishinaga; **p. 171** EMPICS /Tony Marshall; **p. 173** Corbis/Royalty-Free; **p. 174** Action Plus/Glyn Kirk; **p. 177** *l* Science Photo Library/Dr Jeremy Burgess; *r* Science Photo Library/Charles D. Winters; **p. 181** Science Photo Library/David Scharf; **p. 183** *c* Corbis/Catherine Karnow, *bl* Science Photo Library/Agstock/ Gary Holscher; **p. 184** Science Photo Librry/Adam Hart-Davis; **p. 185** *l* PurestockX, *r* PurestockX; **p. 186** *l* FLPA/Holt Studios/Bob Gibbons, *r* Science Photo Library/Paul Rapson; **p. 187** *t* ABPL/Amos Schliak/More Images, *tl* ABPL/Amos Schliak/More Images, *cl* PurestockX, *bl* PurestockX; **p. 189** *l* Science Photo Library/Andrew Syred, *r* Science Photo Library/Maximilian Stock Ltd; **p. 190** *cl* Science Photo Library/Rosenfeld Images Ltd, *bl* Science Photo Library/Brian Bell, *br* Alamy/Phototake Inc.; **p. 197** Science Photo Library/Russell Kightley; **p. 198** Science Photo Library/Hank Morgan; **p. 199** Alamy/Phototake Inc.; **p. 200** Science Photo Library/Prof. David Hall; **p. 202** Ashden Awards for Sustainable Energy www.ashdenawards.org/ Martin Wright; **p. 203** Ashden Awards for Sustainable Energy www.ashdenawards.org/ David Fulford; **p. 204** Science Photo Library/Martin Bond; **p. 205** Rex Features/Jerry Daws; **p. 207** Rex Features/Jason Bye; **p. 208** *tl* Science Photo Library/Andrew Mcclenaghan, *br* Science Photo Library/Maximilian Stock Ltd; **p. 210** Science Photo Library/Cordelia Molloy.

b = bottom, *c* = centre, *l* = left, *r* = right, *t* = top

Every effort has been made to trace all copyright holders, but if any have been inadvertently overlooked the Publishers will be pleased to make the necessary arrangements at the first opportunity.

Risk assessment
As a service to users, a risk assessment for this text has been carried out by CLEAPSS and is available on request to the Publishers. However, the Publishers accept no legal responsibility on any issue arising from this risk assessment: whilst every effort has been made to check the instructions for practical work in this book, it is still the duty and legal obligation of schools to carry out their own risk assessment.

Hodder Headline's policy is to use papers that are natural, renewable and recyclable products and made from wood grown in sustainable forests. The logging and manufacturing processes are expected to conform to the environmental regulations of the country of origin.

Orders: please contact Bookpoint Ltd, 130 Milton Park, Abingdon, Oxon OX14 4SB.
Telephone: (44) 01235 827720. Fax: (44) 01235 400454. Lines are open 9am–5pm, Monday to Saturday, with a 24-hour message answering service. Visit our website at www.hoddereducation.co.uk

First published in 2007 by
Hodder Murray, an imprint of Hodder Education,
a member of the Hodder Headline Group
338 Euston Road
London NW1 3BH

Impression number 5 4 3 2 1
Year 2011 2010 2009 2008 2007

Cover photos Dragonfly: Andy Harmer/Science Photo Library; Crop research: Maximilan Stock Ltd/Science Photo Library; Chromosomes: Dept. of Clinical Cytogenetics, Addenbrookes Hospital/Science Photo Library
Illustrations by Barking Dog Art
Typeset in Times 11.5pt by Fakenham Photosetting Limited, Fakenham, Norfolk

Printed in Italy

A catalogue record for this title is available from the British Library.

ISBN: 978 0 340 92797 7

Contents

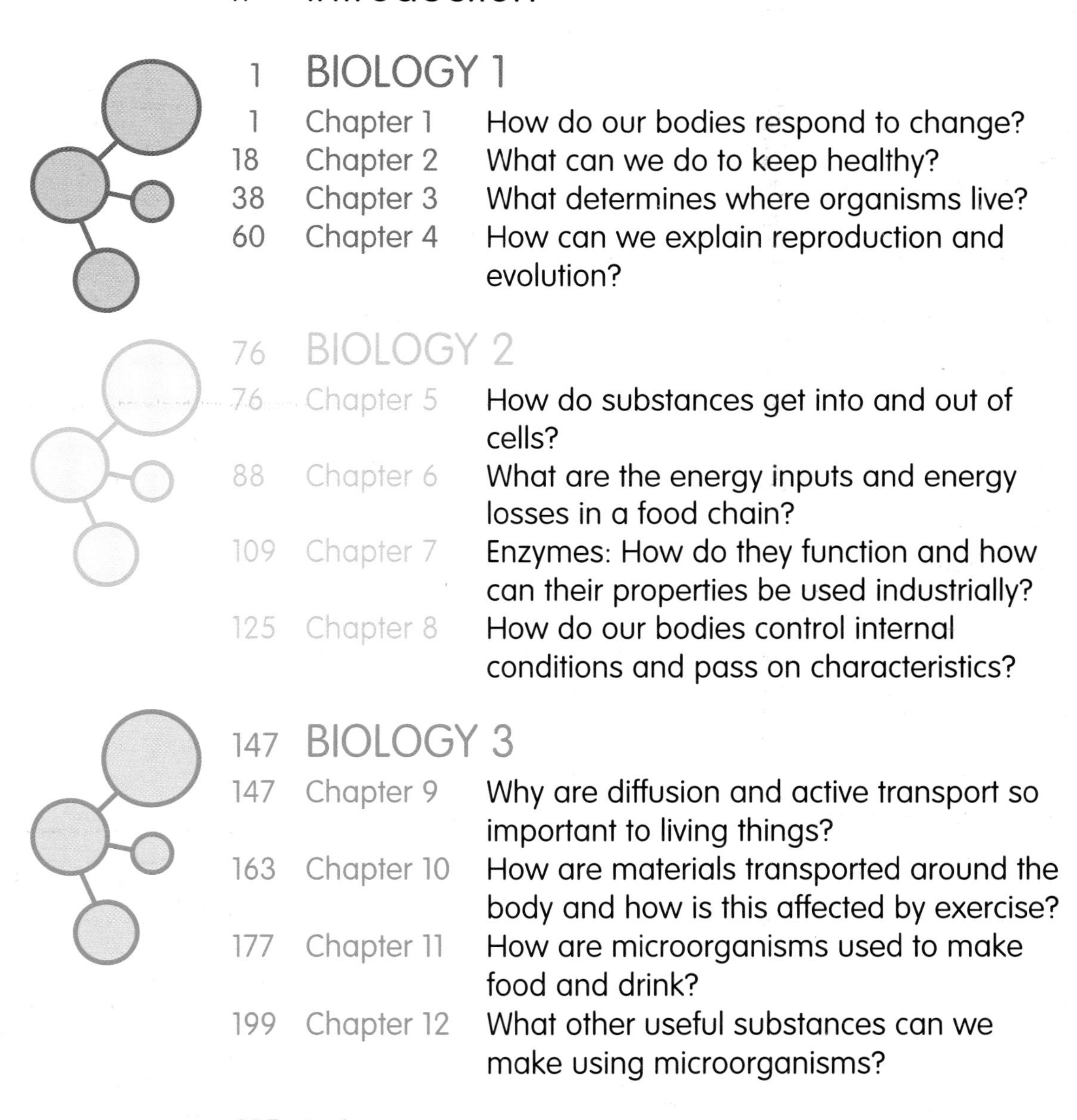

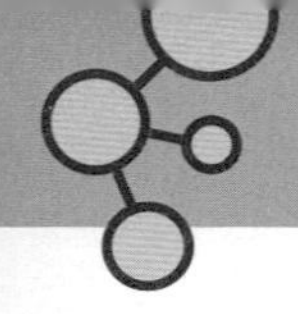

Introduction

Welcome to the AQA GCSE Biology Student's Book. The book covers all the Biology content as well as the key 'How science works' elements.

Each chapter starts with a set of **learning objectives**. Don't forget to refer back to these when checking whether you have understood the material covered in a particular chapter. **Questions** appear throughout each chapter, which will help to test your knowledge and understanding of the subject, as you go along. They will also help you develop key skills and understand how science works.

Activities are found throughout the book.* These will take you longer to complete than the questions but will show you many of the real-life applications and implications of science. At the end of each chapter a **summary** evaluates the important points and key words. You will find the summaries useful in reviewing the work you have completed and in revising for your examinations. Don't forget to use the **index** to help you find the topic you are working on.

You will find **exam-style questions** at the end of each chapter, to help you prepare for your exams. These include similar questions to those in the unit tests.

You will sit three written papers for GCSE Biology: Biology 1, Biology 2 and Biology 3. These match the three sections of this book.

This book includes both **higher-tier** and **foundation-tier** material. Learning objectives and summary points needed for the higher tier only are shown by a tick in a green circle (✓). In the text, the sections of the book for higher tier only are shown with a thick green stripe along the right-hand side of the page. Questions numbered inside a green circle (e.g. 2) would be asked only in the higher-tier exam.

Finally, we would like to thank Gillian Lindsey, Becca Law and Chris Wyard, members of the Science Team at Hodder Murray, for their helpful and perceptive contributions to the production of this book.

Good luck with your studies!

Toby Houghton and Christine Woodward

* The activities in these chapters involve minimal risk if normal school laboratory rules are adhered to. However, some teachers may wish to amplify their lessons with additional traditional or novel experiments and / or demonstrations that use more hazardous materials, for example cheek cell sampling, measuring blood pressure or testing antibiotics. For these experiments, a Risk Assessment is required from the employer. All Local Authorities and the majority of independent schools use the services of CLEAPSS for advice on Risk Assessments. Subscribers to CLEAPSS will find useful advice in the CLEAPSS Laboratory Handbook, which is on the CLEAPSS CD-ROM, or they could contact CLEAPSS directly on the *Helpline*.

Chapter 1
How do our bodies respond to change?

At the end of this chapter you should:

- ✓ know how the different parts of the nervous system work together to co-ordinate our response to external stimuli;
- ✓ be able to explain the difference between a voluntary and a reflex response;
- ✓ understand that many processes in the body are co-ordinated by chemicals called hormones;
- ✓ be able to explain how hormones control the amount of water and sugar in the blood;
- ✓ be able to describe how hormones control the menstrual cycle;
- ✓ know how hormones can be used to control a woman's fertility;
- ✓ be able to explain how medical drugs are developed and tested;
- ✓ be able to describe the effects of a range of drugs on the body;
- ✓ be able to evaluate the claims of researchers about the potential risks of cannabis and its links with addiction to hard drugs.

Figure 1.1 In each of these pictures people's bodies will respond to what is happening

How do different parts of our nervous system work together and respond to change?

Look at the pictures in Figure 1.1.

1. Describe how you think each person will respond to the situations shown in the pictures.
2. For each different scenario, explain how you think the body controls its response.

Receptors are cells that can detect changes inside or outside the body.

The central nervous system (CNS) is the spinal cord and the brain.

Effectors are organs in the body that cause a response. They can be muscles or glands.

Your body is constantly responding to changes taking place outside and inside your body. If someone throws you a ball you respond quickly by moving your hands to catch it. If a teacher is about to ask you a question in class you may feel nervous. If you haven't drunk enough water on a hot day, your body will make changes to save the water you already have in your blood.

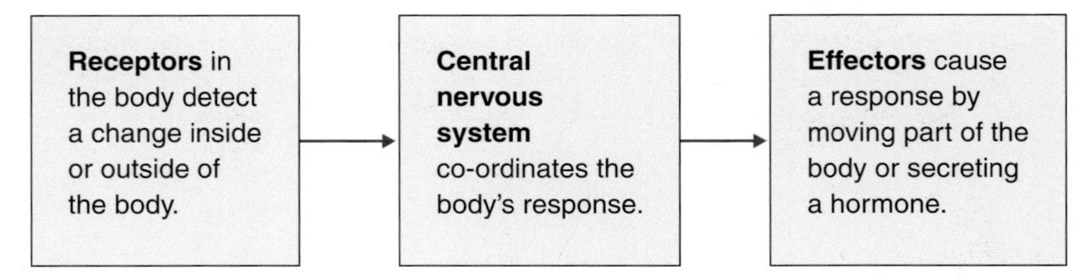

Figure 1.2 A flow diagram showing how the nervous system enables the body to respond to changes

For example, if a friend calls your name, receptors in your ears detect the sound. They then send a message to your **central nervous system** (CNS). Your CNS registers that you need to turn round and so sends a message to your muscles. The muscles (the **effectors**) then contract so that you are facing your friend.

In your body, receptors are found in the **sense organs**. Eyes are the sense organs that contain light receptors. The changes that receptors detect are called **stimuli**. For example, light is the stimulus detected by the light receptors in your eyes. When a receptor detects a stimulus, it converts it into a tiny electrical impulse.

3. List four changes, either internal or external, that your body can respond to.
4. For each of the changes you have listed, write down where the receptors to detect this change are found and what the response might be. You could put your answers into a table.

Sense organs are organs that contain receptor cells.

Stimuli (singular: stimulus) are changes that receptor cells detect.

There are eight different types of receptors in your body. These are shown in Table 1.1.

Stimuli	Receptor	Sense organ
Light	Light receptors	Eye
Sound	Sound receptors	Ear
Movement	Position receptors Touch receptors Pressure receptors	Ear Skin Skin
Chemicals	Chemical receptors	Tongue Nose Blood vessels
Change in temperature	Temperature receptors	Skin Blood vessels
Pressure/temperature	Pain receptors	Skin

Table 1.1 The range of receptors found in the body and the stimuli they detect

Figure 1.4 The people in each of these pictures are about to respond to stimuli they can detect.

You should now be able to explain how the nervous system responds to a range of stimuli. A good way to do this is to draw a flow diagram describing each stage. Figure 1.3 shows how the nervous system responds when you touch a hot object, like a pan on a stove.

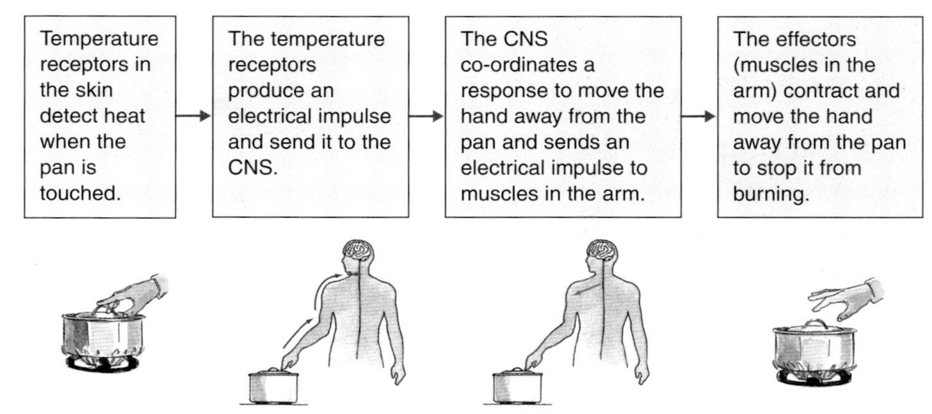

Figure 1.3 A flow diagram showing how the nervous system responds so that a person will not burn themselves seriously when they touch a hot object.

5 Draw flow diagrams showing how the nervous system co-ordinates a response to each situation in the three pictures in Figure 1.4. Try to include as much detail as that given in the example in Figure 1.3.

How do electrical impulses get round your body?

If you call friends who live two miles away, the phone lines connect you with your friends and carry your message to their phone. Your body connects the receptors, CNS and effectors together in a similar way. In your body, the **nerves** act like the wires in the phone lines. They conduct the electrical impulses from one place to another.

Nerves are made up of individual nerve cells called **neurones**. Neurones are laid end to end to make long strands, a bit like individual wires. These long strands are bundled together to make nerves. An electrical impulse starts at one end of a neurone and passes along it to the other end.

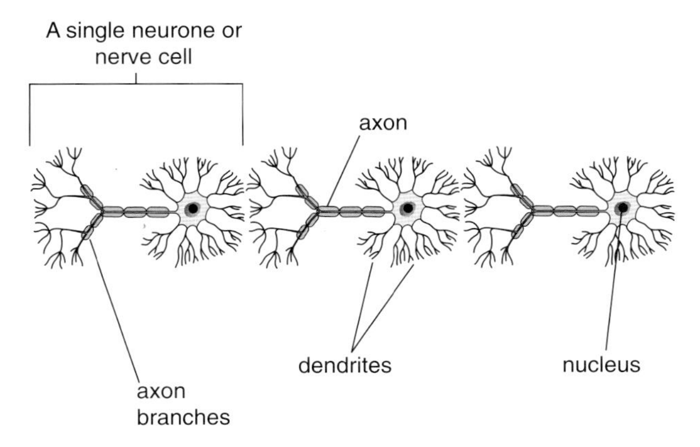

Figure 1.5 Three neurones forming a single strand in a nerve

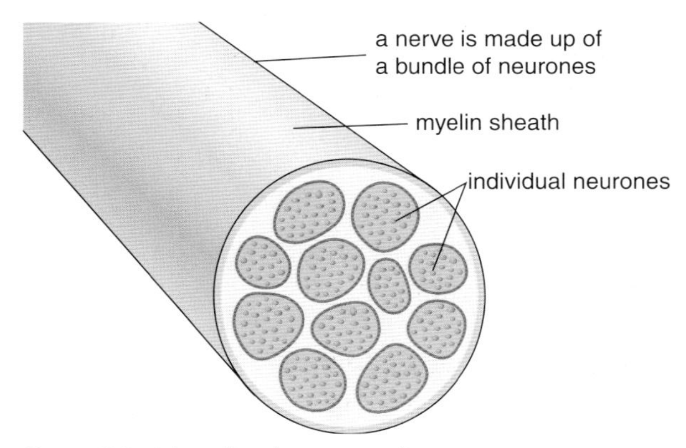

Figure 1.6 A bundle of neurones forms a nerve

Neurones are individual nerve cells that are specialised to transmit electrical impulses.

Nerves are bundles of neurones that connect receptors and effectors to the CNS.

A synapse is the gap between the ends of two neurones.

Chemicals that pass across a synapse and cause an electrical impulse to be generated in the next neurone are called **chemical transmitters**.

6. How is a neurone different to other types of animal cells?
7. How do these differences ensure a neurone is effective at transmitting electrical (nerve) impulses around the body?
8. Describe how information is sent from a temperature receptor in your finger to your brain.

Voluntary actions are responses that are co-ordinated by the brain.

Reflex actions are automatic, rapid responses, often to harmful situations.

Sensory neurones carry impulses from a receptor to the spinal cord.

Relay neurones carry impulses from sensory neurones to motor neurones.

Motor neurones carry impulses to effectors.

Between the end of one neurone and the start of another, there is a gap called a **synapse**. The electrical impulse cannot travel across this gap, but it causes the formation of a **chemical transmitter** which triggers an electrical impulse in the next neurone. Once the new impulse has been sent, the chemical is broken down by an enzyme. This process can happen between every neurone in your body and this ensures that the signals can move to all parts of your body.

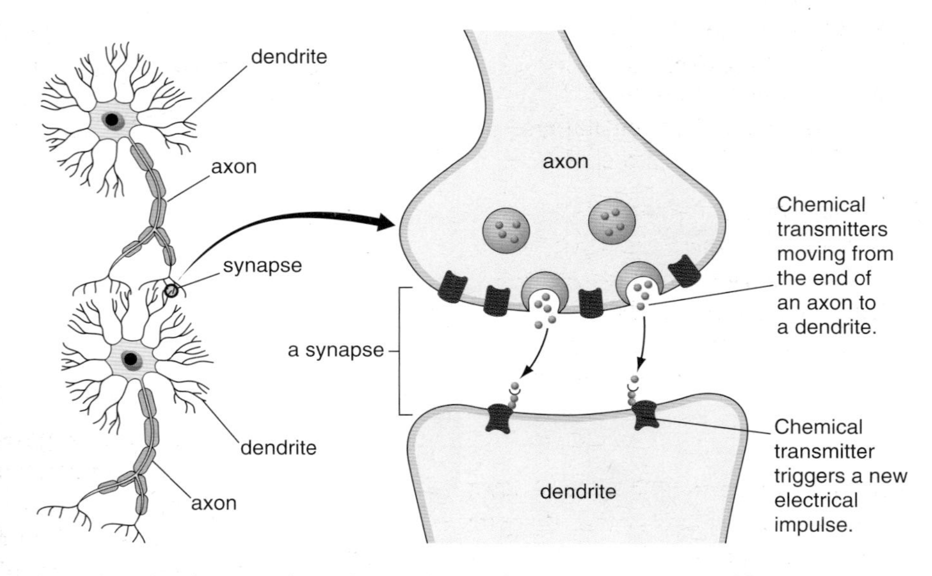

Figure 1.7 This diagram shows how a chemical transmitter is released from one neurone. This chemical travels across the synapse and then triggers the release of a new electrical impulse in the next neurone.

How fast can the nervous system respond to stimuli?

Electrical impulses can travel along your neurones at about 100 metres per second. That's the speed you would need to run to finish a 100 m race in 1 second! So, all responses are fast but for many responses you need to add the time it takes for the brain to process the information and co-ordinate a response. These responses that are co-ordinated by the brain are called **voluntary actions**.

Voluntary actions are fine if you are responding to a question from your friend, but if you need to respond to a harmful situation you need to respond more rapidly. For example, if you touch a sharp object, you pull your hand away before thinking about it. This type of automatic response is called a **reflex action**.

Reflex actions do not rely on the brain to co-ordinate a response. This is brought about by the spinal cord. A **sensory neurone** carries the impulse to the spinal cord where a **relay neurone** passes the impulse straight to a **motor neurone**. The motor neurone then carries the impulse to the effector to trigger a response.

9. Which of the responses listed below are voluntary actions and which are reflex actions?
   - Blinking when someone kicks dust at you.
   - Sneezing when pepper goes up your nose.
   - Picking up a pen.
   - Pulling your hand away when you touch a hot iron.
   - Turning around when someone calls your name.
   - Taking a CD off a shelf.
10. Choose one of the reflex actions from question 9. Draw a flow diagram like Figure 1.3 to show how the body co-ordinates this response.

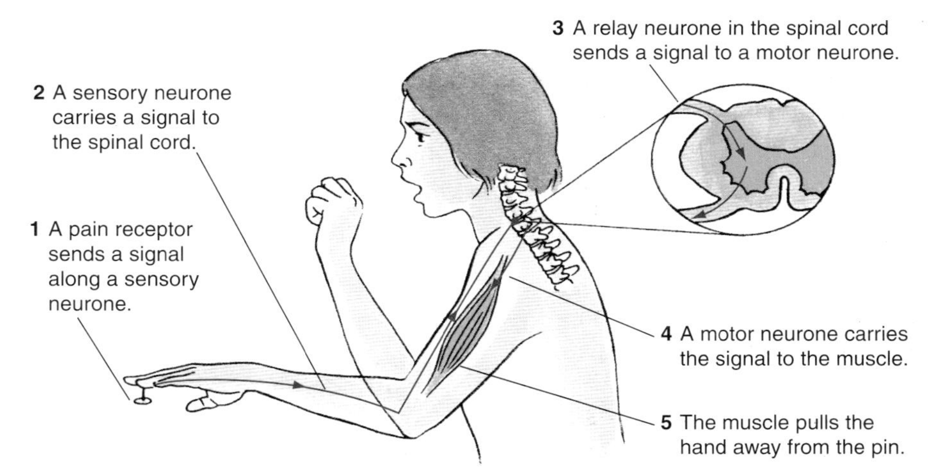

Figure 1.8 A reflex action brings about a quick response when someone touches a sharp object

Activity – Multiple sclerosis

Multiple sclerosis (MS) is a disease that damages the central nervous system. Its symptoms are very varied and can include: blindness, slurred speech, poor co-ordination, paralysis and memory loss. At present there is no cure for multiple sclerosis, but there is a lot of research being carried out to develop treatments for the disease.

The nerves in the central nervous system are coated with a chemical called myelin. This insulates the nerve in a similar way to the plastic coating on electrical wire. When a person suffers from MS, patches of myelin become damaged, exposing the nerve itself.

1. What are the symptoms of MS?
2. From what you have learnt about the nervous system, suggest how a damaged myelin coating can cause the symptoms of MS.

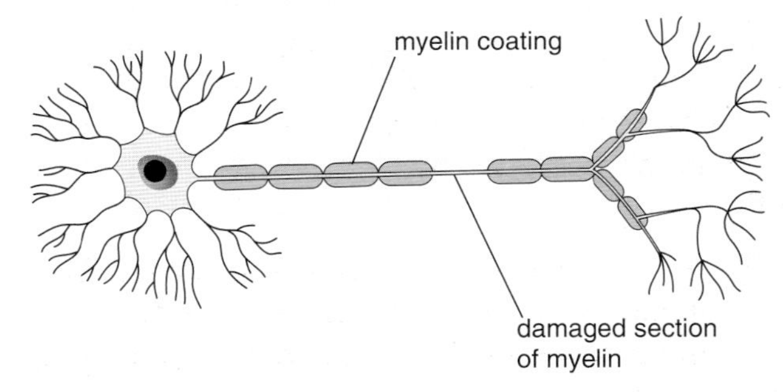

Figure 1.9 A nerve with a damaged myelin coating in a person suffering from MS

3. MS only causes damage to nerves in the brain and spinal cord. However a sufferer may have poor co-ordination of their legs. Why does this happen if the nerves in the legs are not actually damaged?
4. Prepare a list of six 'frequently asked questions', with answers, that could be published on a website for people who have recently been diagnosed with MS. Think about the information that would really help. Try using the following websites to help you: www.mstrust.org.uk www.mult-sclerosis.org
5. Scientists are constantly carrying out research. The increase in our scientific knowledge has helped us to answer many questions. But science cannot supply all the answers. Look carefully at the six 'frequently asked questions' that you have listed in question 4.
   a) Which questions do you think can be answered using our present scientific knowledge?
   b) Which questions are outside the boundaries of science? (Perhaps they are questions where beliefs, opinions or personal views are important.)

1.2 How do hormones control conditions inside your body?

Hormones are chemicals which are released from glands into the blood and then transported to their target organs.

Glands are the parts of the body that release (secrete) hormones.

A target organ is the organ that a specific hormone acts upon.

When asked about **hormones**, people often mention teenage mood swings and puberty. Both of these changes are triggered by hormones. This topic will help you to explain how hormones control lots of other processes in your body. If something scares you, or puts you under stress, your heart beats faster. This happens because a chemical called adrenaline is released into your bloodstream and then travels to your brain. Your brain responds by sending a nerve impulse to your heart, causing it to beat faster. Adrenaline is one of the many hormones produced by your body. Organs that release hormones are called **glands**. They act on specific organs called **target organs**.

How does a hormone control the water content of the body?

It is essential to have the right amount of water in your body. You take in water when you eat and drink but water is lost from your body in a number of ways. Water is lost when you breathe out, when you sweat and water leaves your body in urine when you go to the toilet.

One way in which your body can control the loss of water is by altering the amount of water in your urine. If you have too little water in the blood, your kidneys will reduce the volume of water used to make urine.

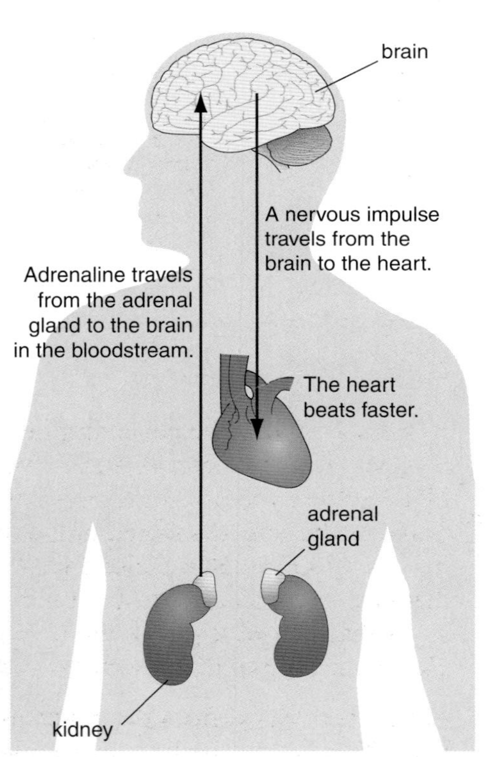

Figure 1.10 How the release of adrenaline can affect the heart

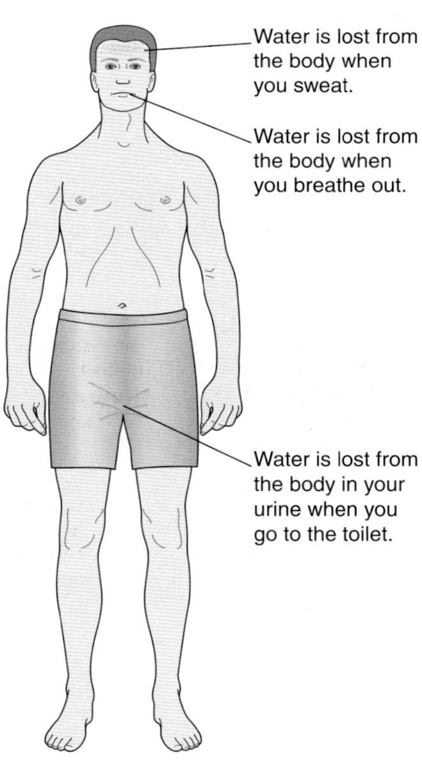

Figure 1.11 Ways in which water is lost from the body

⓫ Look at Figure 1.10. Name the gland, hormone and target organ in this response.

⓬ Compare the response system in question 11 with the way goods are made, transported and delivered in everyday life.

13. State two conditions that might cause the amount of water in your blood to decrease.
14. What happens to the level of ADH in the blood when the amount of water in your blood decreases? Explain your answer.
15. What happens to the concentration of your urine when you drink a) lots of water; b) too little water? How are these changes controlled?

The water that is saved goes back into the blood. If you have too much water in the blood, your kidneys allow the excess to be lost in urine. This response in the kidneys is controlled by anti-diuretic hormone (ADH), which is secreted by the pituitary gland. ADH is carried by the blood to the kidneys, its target organs. An increase in ADH causes the kidneys to re-absorb more water into the blood. A decrease in ADH causes the kidneys to leave more water in the urine.

When water is lost from the body, in sweat and urine, essential ions such as sodium (Na^+) and potassium (K^+) are also lost. So when your body controls the amount of water in the blood it is also controlling the ion content of your body.

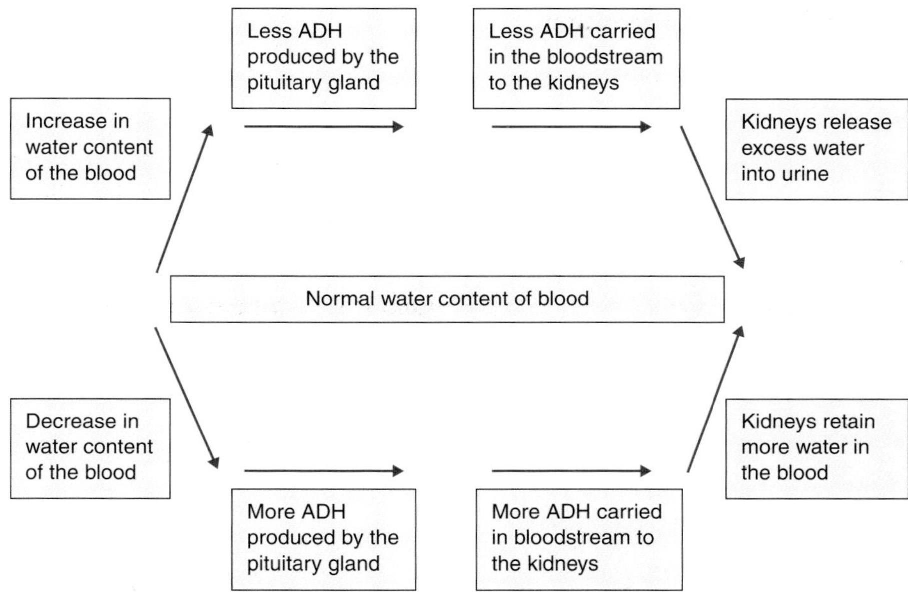

Figure 1.12 A flow chart summarising the control of water content in the blood by ADH and the kidneys

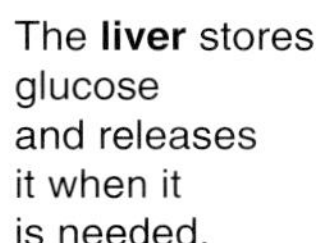

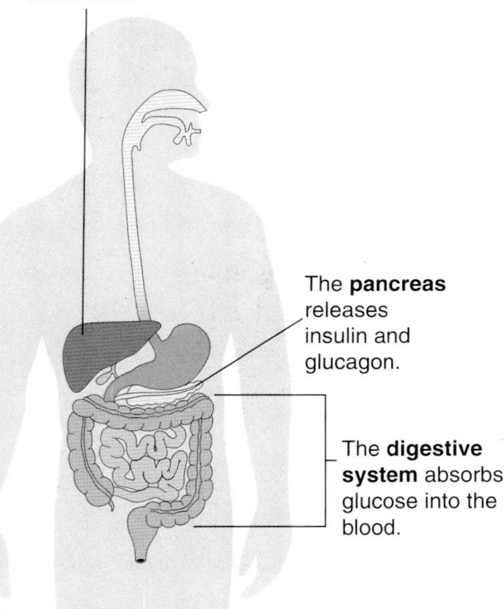

Figure 1.13 The pancreas where insulin and glucagon are released and the liver where sugar is stored

How do two hormones work together to control the concentration of sugar in your blood?

It is essential that the concentration of sugar in your blood is controlled. It should stay between 90 and 100 mg of sugar in every 100 cm^3 of blood. When you eat, the amount of sugar in your blood increases. Some of this sugar is stored in the liver. It can then be released back into the blood when the blood sugar level drops. The amount of sugar that is stored is controlled by a hormone called **insulin**. The amount of sugar that is released from the liver is controlled by a hormone called glucagon. Both of these hormones are released from the pancreas.

Insulin is released from the pancreas when blood sugar concentration rises. This causes the liver to remove sugar from the blood and store it. When blood sugar drops too low, glucagon is released by the pancreas. This causes the liver to release stored sugar back into the blood.

16. Name a) the gland; b) the target organ; c) the hormones involved in the control of blood sugar concentration.

17. A patient's blood sugar concentration was measured six times during a five hour period. The following values were obtained in mg of sugar per 100 cm^3 of blood:

92, 89, 95, 97, 115, 97.

Copy and complete the following sentences.

The **range of data** for the patient's blood sugar concentration varies from __________ (the lowest value) to __________ mg of sugar per 100 cm^3 of blood (the highest value).

One of the values appears to be **anomalous**. The anomalous value is __________ mg of sugar per 100 cm^3 of blood.

Ignoring the anomalous result, the patient's mean (average) blood sugar concentration was __________ mg sugar per 100 cm^3 of blood.

A collection of measurements is called **data** (singular **datum**). The **range of data** is from the smallest to the highest value.

Anomalous results are very different from the others and do not lie within the expected range.

Evidence is data that have been subjected to some form of testing or validation.

How do hormones control the menstrual cycle in women?

A woman's menstrual cycle lasts about 28 days. During the cycle the lining of the womb (uterus) thickens. An egg matures and is released from the ovaries. If this egg is fertilised by a sperm cell it may implant in the lining of the womb. If the egg does not implant in the womb much of the lining is then shed during the woman's period. All these changes are controlled by hormones released by the ovaries and the pituitary gland in the brain (Table 1.2). The levels of these hormones in the bloodstream change throughout the menstrual cycle.

Hormone	Released by	Effect on the body
Folicle stimulating hormone (FSH)	The pituitary gland	• Causes an egg to mature in an ovary • Stimulates the ovaries to produce oestrogen
Luteinising hormone (LH)	The pituitary gland	• Stimulates the release of an egg from an ovary
Oestrogen	The ovaries	• Inhibits further production of FSH • Stimulates the pituitary gland to release LH • Causes the lining of the womb to repair and thicken

Table 1.2 The main hormones involved in control of the menstrual cycle

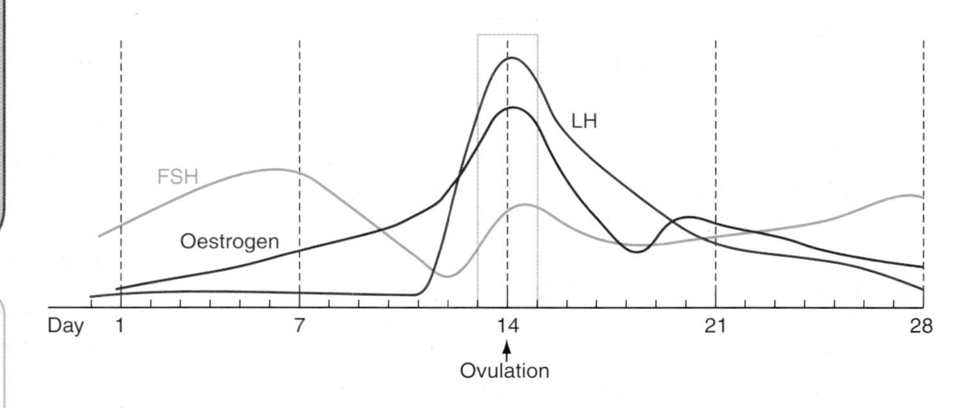

Figure 1.14 A graph showing how the levels of hormones change during the menstrual cycle

18. Sketch a copy of the graph in Figure 1.14. Add labels to it to describe the changes caused by each hormone.

Activity – Using hormones to control fertility

The hormones involved in the menstrual cycle can be used to help a woman become pregnant or to prevent this happening (contraception). Hormones in the contraceptive pill reduce fertility and prevent a woman from becoming pregnant.

Laura and David have just got married. They plan to have children in about three years' time but until then they want to have sex without Laura getting pregnant. Laura talked to her doctor about possible methods of contraception. She suggested that Laura should go on the pill and explained that the pill contained a hormone that would prevent an egg maturing in the ovaries each month.

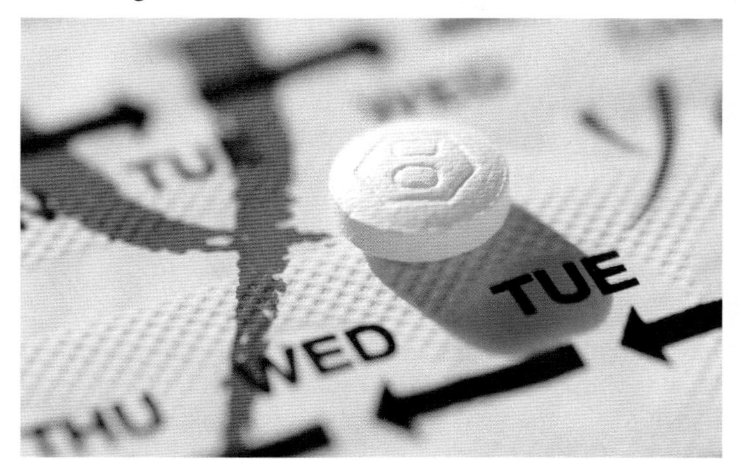

Figure 1.15 The contraceptive pill Laura was prescribed

1. Why do you think Laura's doctor suggested the pill as a method of contraception? Discuss this in a pair and write a list of reasons for using the pill.
2. What possible disadvantages are there in using the pill as a method of contraception? Use the following website to help you answer this question: www.fpa.org.uk. Click on the following links: 'Contraception and sexual health guide', 'Contraception', 'The combined pill', 'Are there any risks?'
3. The pill contains the hormone oestrogen. Use Table 1.2 to explain how oestrogen can prevent an egg from maturing.

After three years Laura and David decided to start a family so Laura stopped taking the pill. They had sex regularly, but after a year Laura was still not pregnant. Her doctor referred them to a fertility clinic. After some tests the doctor found that the infertility was due to Laura's eggs not maturing properly in the ovaries. This was caused by a low concentration of one of her hormones. She was prescribed a fertility drug that contained a synthetic form of the hormone. The hormone caused her eggs to mature properly in the ovaries. After six months Laura became pregnant.

4. Look at Table 1.2. What hormone do you think the fertility drug contained? Explain your answer.
5. a) Some people do not agree with the use of synthetic hormones to treat infertility. Their reasons may be based on:
   A evidence, B hearsay, C prejudice,
   D personal opinion.

   Look carefully at the reasons given by the following four people for not agreeing with the use of synthetic hormones to treat infertility. In each case, decide which of A, B, C or D their reason is based on.
   - i) Kelly says 'It's wrong to interfere with nature.'
   - ii) Ali, her husband says 'The use of synthetic hormones could create problems in Kelly's hormonal system.'
   - iii) Becky says 'It's against my religious beliefs.'
   - iv) Lisa, her friend, says 'The synthetic hormones are drugs and all drugs are bad for you.'

   b) What is your opinion on the use of synthetic hormones to treat infertility?

In some cases of infertility taking this hormone does not help a couple have children. Another treatment is called IVF (*in vitro* fertilisation). This involves the woman taking another hormone to stimulate egg production. Eggs are then removed from her ovaries with a needle and fertilised with the man's sperm in a laboratory. A few of the fertilised eggs are then inserted into the woman's womb in the hope that one or more of them will implant and develop into a baby.

6. What causes of infertility do you think IVF can be used to treat?
7. Suggest two potential problems related to IVF.

1.3 How do our bodies maintain the right temperature?

Internal body temperature in °C	Symptoms
28	Muscle failure
30	Loss of body temperature control
33	Loss of consciousness
37	Normal
42	Central nervous system breakdown
44	Death

Table 1.3 Symptoms that occur when the body's internal temperature changes

Figure 1.16 A marathon runner on a hot day

19 Look at the photo in Figure 1.16. How is the marathon runner's body responding to the increase in her body temperature?

20 How do you think these changes reduce her body temperature?

Look at Table 1.3. It shows that changes in your body temperature can have very serious effects on how it functions. Most of these changes take place because enzymes controlling the processes in your body work best at normal body temperature. This is 37 °C. When the temperature changes, enzyme molecules begin to change shape. If this happens they cannot catalyse body processes effectively.

Fortunately your body has a number of ways of controlling its temperature. It can cool itself down if it gets too hot and warm up if it gets too cold.

How does the body cool down?

Your body will warm up on a hot day or when you take exercise. Imagine a marathon runner on a hot day and you can probably picture how the body responds to its increasing temperature.

Sweating is often the first sign that your body is trying to reduce its internal temperature. Sweat is produced by sweat glands in the skin and secreted onto its surface. As water in the sweat evaporates off the skin, it takes heat away from the body. This reduces the internal body temperature.

When your skin becomes red on a hot day, it has more blood flowing near the surface. This happens because your brain sends a nervous impulse to the blood capillaries near the surface of your skin which causes them to dilate (widen). Some of the heat that the blood carries to the surface is then lost to the air. This response is called vasodilatation (see page 128).

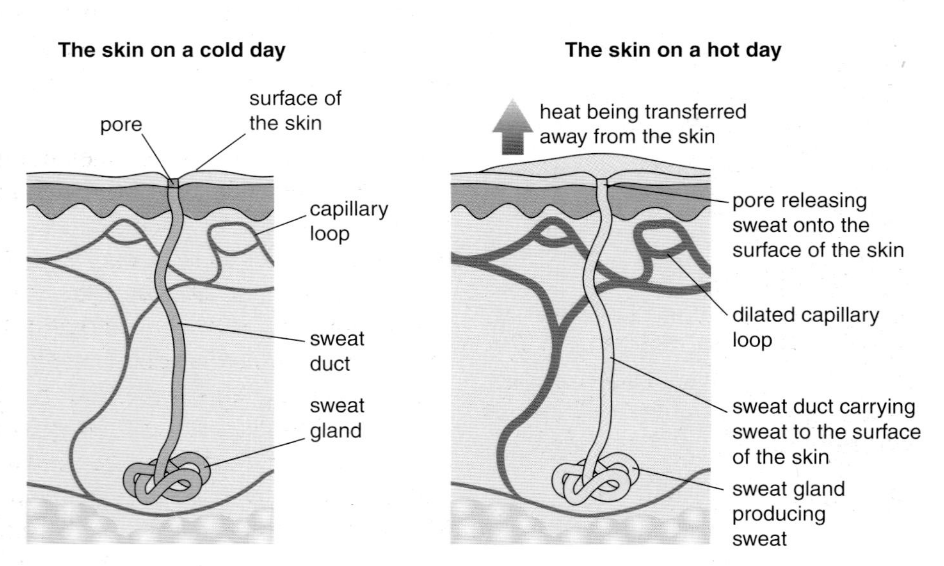

Figure 1.17 Changes that occur on the surface of the skin on a hot day

21. Explain how stopping sweating, vasoconstriction and shivering help to raise body temperature.

22. Hypothermia is a condition that occurs if body temperature drops below 35 °C. Many more old people than young people die from hypothermia. Why do you think this is? Explain your answer.

How does the body warm up?

When the temperature around you drops, your body temperature will start to fall. Your body will respond to this by retaining the heat that is being released during respiration. It does this by stopping sweating and by reducing the blood flow through the capillaries in the surface of the skin. This happens when the capillary loops in the skin (Figure 1.17) are closed off by a process called vasoconstriction. Your body will also produce more heat by shivering. Shivering causes your muscles to contract and relax very quickly. This causes respiration in your muscle cells and heat is produced.

1.4 How do drugs affect the body?

Figure 1.18 A range of medical drugs available today

Drugs are substances that have an effect on processes in the body. Medicines are drugs. When used properly, they are beneficial to the body. Some people use substances like nicotine, alcohol, cannabis and heroin as recreational drugs. Some of these are more harmful than others. It is very important to know exactly how a drug affects our bodies before we can assess its possible risks or benefits.

Medical drugs

Medical drugs or medicines are used to treat disease, injury and pain. Medicines are beneficial when used properly but can still be harmful if misused. Paracetamol is an effective painkiller if you take the correct dose but taking more than this can cause liver damage. In extreme cases, it can kill. This is why it is essential that you always follow the instructions given with medicines.

There is evidence of medicines being used as far back as prehistoric times (8000BC). Cave paintings show tribal healers feeding people plants to cure illness. The Egyptians had doctors in 200BC who used herbal medicines to treat diseases. Some of the medicines used today contain chemicals that were present in these ancient treatments and many more are derived from natural substances found in plants today. Over 7000 chemicals used to make medicines are found in plants and many of these plants are found in the rainforests. The indigenous people of the rainforest use the plants growing there to treat many illnesses. Quinine is extracted from the cinchona tree, found in South America, and is used to treat malaria. A plant called rosy periwinkle, found in Madagascar, provides two of the most important chemicals used to treat cancer. New drugs are still being developed from natural substances found in plants.

a) b)

Figure 1.19 The cinchona tree a) and rosy periwinkle b) are both used to produce essential medicines

23. Why is it important to carefully follow the instructions when you take medicines?
24. Why do you think it is important to medical science that the rainforests are preserved?
25. What are the advantages of testing a new drug on live cells instead of on animals?
26. Why do you think new drugs are tested on healthy people instead of people with the disease in stage 2 of the testing process?
27. Why do scientists bother testing a new drug against an existing medicine in stage 4 of the process?
28. Describe the side effects of the drug thalidomide. Why do you think the side effects of thalidomide were not detected when the drug was tested?

Figure 1.20 A person suffering from the side effects of thalidomide

When scientists develop a new drug to treat a disease it must be thoroughly tested. The testing is carried out in a number of stages.

Stage 1: Drugs are tested in the laboratory. They are tested on live cells and sometimes on animals.

Stage 2: Drugs that show promise in the laboratory are tested on a small number of healthy humans.

Stage 3: The drug is tested on about 100 humans who suffer from the disease that the drug is intended to treat.

Stage 4: The drug is tested on several hundreds of sufferers alongside an existing drug for the disease.

Only drugs that pass all safety checks and clinical trials can be launched as a new product. In the vast majority of cases medical drugs, when used as instructed, have no, or only limited, side effects. Very occasionally a drug that has been tested causes unexpected side effects. A drug called thalidomide was used in the 1960s as a sleeping pill. It was also effective at relieving morning sickness in pregnant women. Unfortunately, many babies born to mothers who took the drug had very underdeveloped arms and legs. The drug was then banned but has since been used successfully to treat leprosy.

The testing and development of modern medicines and drugs illustrates the way in which research scientists work. The process often begins with an idea or an explanation using existing theories as to why a particular drug or medicine might be used to treat a specific disease. This idea or explanation is called a **hypothesis**. The hypothesis can then be used to make **predictions** about the effects of the drug which can be thoroughly tested.

The tests involve laboratory experiments before any trials on humans.

If the experiments and observations support the initial hypothesis about the drug, then clinical trials can begin. These trials will use carefully selected groups to check the effects of the drug on people of different sex, age and weight. They will also involve control groups to ensure that the effects of *only the drug* (the **independent variable**) are observed. The control group will be given a placebo (tablets that look like the drug but have no effect). All those involved in the clinical trials will not be told whether they are having the experimental drug or the placebo.

If, after all these tests and trials, the results support the initial predictions about the drug, then large-scale production can begin.

Activity – What does research tell us about the use of cannabis?

Recently there has been a lot of research into the use of cannabis. One area of research has investigated the possible link between taking cannabis and then moving on to hard drugs, such as heroin. Another area of study has tried to establish a link between smoking cannabis and health problems, especially psychological problems. There have been some conflicting findings.

Read the following quotes from different researchers.

i) 'Cannabis use does not lead to experimentation with harder drugs.'
ii) 'Cannabis does not act as a "gateway" drug to the use of harder drugs.'
iii) 'Many hard drug users have followed a similar path from cigarettes and alcohol, to cannabis, to heroin and cocaine.'
iv) 'Early marijuana smokers were found to be up to five times more likely to move to harder drugs than were their twins who did not smoke marijuana.'
v) 'Some studies have suggested long-term cannabis use can increase your risk of developing schizophrenia.'
vi) 'Smoking cannabis virtually doubles the risk of developing mental illnesses.'
vii) 'There was no proven causal link between taking cannabis and mental illness.'

1. Find two pairs of totally opposite conclusions in the above quotes.
2. Summarise the views expressed in the quotes in four sentences.
3. Different researchers have clearly come to very different conclusions about the links between cannabis smoking, the use of hard drugs and mental illness. In a pair discuss why you think this is. Write a list of the reasons why researchers can come to such conflicting conclusions.
4. Newspapers often use quotes from researchers in their headlines.
   a) What problems might this lead to given the range of opinions expressed by these quotes?
   b) What other information about research into cannabis smoking should be included in a newspaper article to give the public a clearer understanding of the issues?
5. Plan a research project to investigate a possible link between cannabis use and mental illness. You will need to consider the following points.
   - Who will be involved?
   - How many people will be involved?
   - How long should the study last?
   - How will you include a control in the study?
   - How will you collect information from the people involved?
   - How will you analyse the results?
   - How will you share your findings with other people?

A **hypothesis** is an idea or explanation based on accepted scientific theories. A good hypothesis should be able to make **predictions** of results that can be tested.

Variables

A variable is anything that is subject to variation. There are different kinds:

The **independent variable** is the quantity that you change. The **dependent variable** is the quantity you measure when the independent variable changes.

Control variables are quantities that are kept the same.

Continuous variables can have any numerical value, e.g. hormone levels. **Discrete variables** have values that are whole numbers, e.g. the number of receptors.

Categoric variables are different types of something, such as different sexes. **Ordered variables** can be ranked in a clear size order, e.g. small, medium or large clothes sizes.

Figure 1.21 Government health warnings on tobacco products

How does smoking tobacco affect the body?

Tobacco smoke contains more than 4000 chemicals, many of which are poisonous and could kill you. Three of the dangerous chemicals in tobacco smoke are nicotine, tar and carbon monoxide. Nicotine is an addictive substance which causes people to become dependent on tobacco smoke. Tar is actually a mixture of many chemicals which cause lung cancer, bronchitis and emphysema. Carbon monoxide reduces the blood's ability to carry oxygen around the body. It can deprive an unborn baby of oxygen whilst in the mother's womb and lead to a lower birth weight. In spite of all these health problems from the smoking of tobacco, cigarette manufacturers were not required to put government health warnings on cigarette packets until 1971.

Have a look at the key dates below, relating to smoking and health.

1951: Dr Richard Doll and Prof Austin Bradford Hill conduct the first large-scale study of a link between smoking and lung cancer.

1954: Dr Doll and his team publish a paper confirming the link.

1957: The British Medical Research Council announces 'a direct causal connection' between smoking and lung cancer.

1962: A report from the Royal College of Physicians concludes that smoking is a cause of lung cancer and bronchitis, and probably contributes to coronary heart disease. The report recommends tougher laws on cigarette sales, cigarette advertising, and smoking in public places.

1965: The British government bans cigarette advertising on television.

1971: Government health warnings are required on all cigarette packets sold in the UK, following an agreement between the government and the tobacco industry.

29 a) How long did it take from the start of the first large-scale study into tobacco smoking until health warnings were put on cigarette packets?
b) Why do you think this took so long?

30 Which of the dated developments do you think would have the greatest effect in persuading a smoker to give up smoking? Explain your answer.

31 Find more information about the diseases a) bronchitis and b) emphysema.
i) What causes the disease?
ii) What are the symptoms?
iii) How is the disease treated?

32. Research three different approaches that people use to stop smoking. For each method a) explain how it helps a smoker to give up smoking, b) suggest an advantage and a disadvantage for the approach.
33. List the short-term and long-term effects of a) alcohol; b) cocaine.
34. Treatment for drug addicts is freely available on the National Health Service.
    a) State three points in support of this policy.
    b) State one point which criticises this policy.
35. Suggest three things that could be done to reduce the number of drug addicts in the UK.

1976: Prof Richard Doll and Richard Peto publish the results of a 20-year study of smokers. They conclude that one in three smokers dies from the habit.

1983: A report from the Royal College of Physicians features passive smoking for the first time. It also asserts that more than 100 000 people die every year in the UK from smoking-related illness.

1988: An independent Scientific Committee on Smoking and Health concludes that non-smokers have a 10–30% higher risk of developing lung cancer if exposed to other people's smoke.

1989: A UK court rules that injury caused by passive smoking can be an industrial accident.

What other drugs may harm the body?

Drug	Classification	Effects on the body
Alcohol	• Illegal for anyone under the age of 18 to buy • Illegal for someone over 18 to supply to an under 18	• Slows down reactions • Feel relaxed • Lose inhibitions • Can become loud and sometimes aggressive • Unconsciousness • Increased blood pressure • Possible liver and brain damage
Solvents (glue, lighter fluid, paint thinners, correcting fluid)	• Illegal for a retailer to sell a solvent to anyone under the age of 18 if they believe it will be used for inhaling to cause intoxication	• Lose inhibitions • Blurred vision • Dizziness • Blackouts • Possible lung, liver and brain damage
Cocaine	• Class A or 'hard' drug. Illegal to possess or sell	• Feeling of well-being and confidence • Increased heart rate and blood pressure • Highly addictive • Depression • Mental health problems
Heroin	• Class A or 'hard' drug. Illegal to possess or sell	• Feeling of well-being • Drowsiness • Blurred vision • Vomiting • Highly addictive • Can cause a coma or death when taken in large amounts

Table 1.4 The classification and effects of some drugs

Summary

- ✓ The **nervous system** enables humans to react to stimuli from their surroundings and co-ordinate their behaviour.
- ✓ The nervous system includes: **receptors**, **neurones**, the spinal cord, the brain and **effectors** (muscles and glands).
- ✓ Receptors detect stimuli which include light, sound, changes in position, chemicals, touch, pressure, pain and temperature.
- ✓ Information from receptors passes along neurones, in the form of nervous impulses, to the brain. The brain co-ordinates the response.
- ✓ **Reflex actions** are automatic and take place very quickly. They involve sensory, relay and motor neurones.
- ✓ Chemicals called **hormones** are involved in controlling conditions inside the body. These conditions include the water content of the body, the ion content of the body, body temperature and blood sugar levels.
- ✓ Hormones are secreted by glands and are transported by the bloodstream to their **target organ**.
- ✓ Several hormones are involved in controlling the menstrual cycle. These hormones include FSH, oestrogen and LH.
- ✓ Hormones can be used to control **fertility**. This includes using oral contraception to stop a woman's eggs maturing and prescribing FSH as a fertility drug to stimulate eggs to mature.
- ✓ **Drugs** are chemicals that have an effect on processes in the body.
- ✓ **Medicines** are drugs that, when taken properly, are beneficial to people. Other drugs can harm the body.
- ✓ Many drugs are derived from natural substances.
- ✓ When new drugs are developed they must be thoroughly tested. New drugs are tested in the laboratory and then on human volunteers.
- ✓ Occasionally drugs that have been tested cause unexpected side effects. Thalidomide is a drug that caused deformed limbs in the children of mothers who took the drug.
- ✓ Some drugs that are used recreationally are very addictive. Heroin, cocaine and nicotine are examples of addictive drugs.
- ✓ Claims have been made about the effects of cannabis on health and a link between taking cannabis and addiction to hard drugs. However, research has led to conflicting conclusions about these claims.
- ✓ The link between smoking tobacco and lung cancer took some time to be accepted.
- ✓ Tobacco smoke contains a number of chemicals that damage health.
- ✓ Alcohol affects the nervous system by slowing down reactions. It does help people to relax but too much can lead to lack of self-control and even unconsciousness. Alcohol can also cause liver and brain damage.

EXAMQUESTIONS

1. A student accidentally touches a hot saucepan. His hand automatically moves away from the pan.

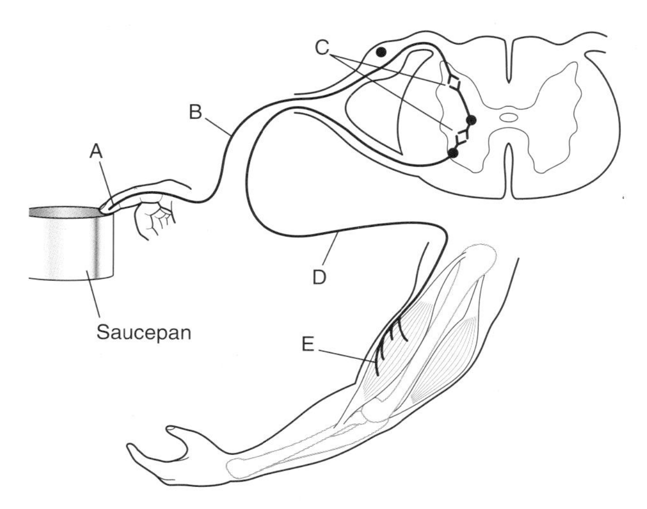

Figure 1.22 The different elements involved in a reflex action

a) In this reflex action:
   i) where is the receptor?
   ii) where is the effector? *(2 marks)*

b) Explain how an impulse crosses the synapse labelled C. *(1 mark)*

c) Explain why this type of reflex action is important to our bodies. *(1 mark)*

2. A hormone is involved in controlling the water content of the body.

a) State **two** ways in which water leaves the body. *(2 marks)*

b) What can cause a rapid fall in the water content of the body? *(2 marks)*

c) Explain how the body responds to a fall in water content. *(3 marks)*

3. Read the information about contraceptive implants.

A contraceptive implant works a bit like a contraceptive pill. They contain a hormone present in contraceptive pills, which prevents pregnancy. The hormone implant is a small, thin flexible rod, 4 cm long and made of plastic. It is inserted just under the skin of a woman's arm. This procedure must always be undertaken by a doctor who is familiar with the technique. The implant gradually releases a small amount of the hormone, which prevents pregnancy for up to three years.

a) How can a hormone prevent pregnancy? *(2 marks)*

b) Give **one** drawback of using contraceptive implants rather than contraceptive pills. *(1 mark)*

c) Contraceptive implants are being used increasingly in birth control programmes in developing countries, instead of contraceptive pills.
Can you think of a reason for this? *(2 marks)*

4. A person's blood sugar level rises after a meal. Following this the pancreas secretes a hormone called insulin. This causes the liver to store sugar until it is needed.

a) i) Name the gland involved in this response. *(1 mark)*
   ii) Name the target organ involved in this response. *(1 mark)*

b) Explain how insulin is transported from the pancreas to the liver. *(2 marks)*

c) Why does the liver need to release sugar when a person is taking vigorous exercise? *(1 mark)*

5. Medicines must be thoroughly tested before they can be prescribed to the public.

a) Outline the procedure for testing a new medicine. *(2 marks)*

b) A drug called thalidomide caused severe side effects despite passing safety tests and clinical trials.
   i) Describe the side effects caused by thalidomide. *(2 marks)*
   ii) Explain why thorough testing did not identify the possibility of these side effects. *(3 marks)*

6. Alcohol and tobacco can damage the body.

a) Give one example of how the body is damaged by tobacco smoke. *(1 mark)*

b) Why do you think that people find it difficult to give up smoking? *(2 marks)*

c) Why should motorists not drive after they have been drinking alcohol? *(2 marks)*

Chapter 2
What can we do to keep healthy?

At the end of this chapter you should:

- ✓ understand the importance of different food groups for your health;
- ✓ be able to discuss the different types of fat and their effect on cholesterol level;
- ✓ understand the link between energy and specific nutrient requirements with age;
- ✓ know how to calculate the percentage of saturated fat in food;
- ✓ appreciate the link between exercise, food energy and health;
- ✓ know how to calculate the amount of energy used with exercise;
- ✓ be able to evaluate whether or not a diet is healthy;
- ✓ be able to explain the body's defences against pathogens;
- ✓ appreciate the importance of immunisation programmes;
- ✓ be able to assess the value of and problems associated with antibiotics.

Figure 2.1 Food can be colourful and interesting

Eating for health

1 a) List eight different coloured fruits or vegetables. Try to include all the colours of the rainbow.
b) Try to name all the fruits and vegetables shown in Figure 2.1.

Do you eat the same foods every day? Do you think about your choice of food? Most people eat a varied diet that includes everything needed to keep their body healthy. The important word in the last sentence is 'varied' because no single food can provide all the essential nutrients your body needs to function efficiently. An easy way to ensure that your diet contains all the essential vitamins and minerals is to make sure you eat as many different naturally coloured foods as possible. Red tomatoes, green broccoli and purple plums are three examples.

For a healthy diet you should aim to eat some food from each of the groups in Table 2.1 every day.

Food group	Examples	Health points
Energy foods	• Whole grains such as rice, millet • Yams, potatoes • Cereals, bread and pasta	• Whole grains provide extra fibre, which means starchy foods (**carbohydrates**) are converted and released as sugars over a longer time period, they are low GI foods. This helps to prevent snacking
Food for building bones and teeth (two or three servings daily)	• Dairy products such as milk, cheese, yoghurt, fromage frais • Soya or rice milk with added calcium	• Skeletons increase in strength most rapidly in the late teens • Skimmed and semi-skimmed milk are as calcium-rich as other milk but have less fats
Muscle-building foods	• Red meat, poultry, sausages, bacon, eggs • Nuts, tofu, beans, lentils, Quorn™ (mycoprotein) • Fish such as salmon, sardines, cod	• Different protein foods are important for vegetarians to ensure they get all the essential **amino acids**. Red meat eaters should avoid too much excess fat • Oily fish such as sardines and salmon provide the **essential fatty acids** (EFAs) which are good for the brain • Meat and fish contain vitamins A, D and E
Foods for clear skin and a healthy digestive system (a minimum of two portions of fruit and three of vegetables daily)	• Leafy vegetables such as cabbage and broccoli • Peas, beans, courgettes, squash, okra, onions • Salad such as lettuce, cucumber, peppers, tomatoes • Fruit such as oranges, nectarines, strawberries, mangoes, blueberries • Dried fruit such as apricots, prunes	• The brightly coloured pigments in fruit and vegetables contain minerals and vitamins • Fruits are rich in vitamin C • The fibre in fruit and vegetables: – speeds the passage of food through the intestine helping prevent colon cancer; – slows the release of sugars, which helps to reduce the risk of diabetes and prevent snacking

Table 2.1 The health points of different food groups

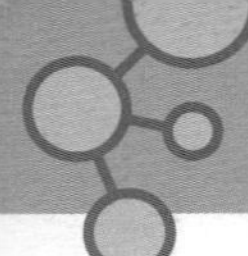

Figure 2.6 A little of what you fancy does you good. Too much makes you fat.

Carbohydrate: a compound containing the elements carbon, hydrogen and oxygen, with atoms of hydrogen and oxygen in the ratio 2:1. Carbohydrates include sugars, starch and cellulose.

Amino acids are the simple molecules (monomers) that form proteins.

Fatty acids form a part of fat molecules.

❷ What health problems might a child start to experience if they refused to eat fruit and vegetables for a few weeks?

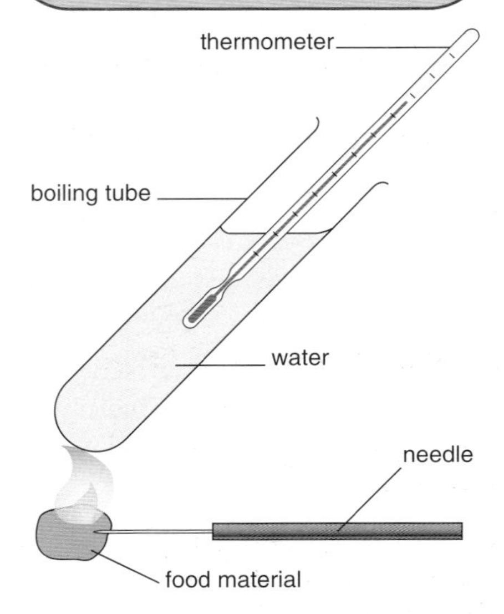

Figure 2.7 Apparatus used to measure the energy value of different foods

If you don't eat all these essential nutrients you will become malnourished whatever your weight. The eating suggestions listed in Table 2.1 contain all the vitamins and minerals you need. But your meals must give you the correct amount of energy or the correct number of joules.

How do we compare the energy content of food?

Energy values are quoted on the 'nutritional information' panels on food packets. In some cases the old units, calories, are still used.

The following experiment will help you to understand how manufacturers obtain the energy value for a food product.

If you were to carry out an experiment to compare the energy values of foods, you might take pieces of food, set fire to them, hold a boiling tube containing water above the flame, and see which heats the water the most.

How many factors have not been controlled in this suggestion and the apparatus shown in Figure 2.7? Think about this before you look at the hints below.

Did you consider the following points?

1. There should be only one independent variable and for this experiment we are trying to compare the energy values of different foods. All other possible variables must stay the same (control variables) for each different food. How could you make the mass of food a **fair test**?
2. Have you tried to burn foods? It can take quite a lot of heat to start the reaction in air. If you heated the foods over a Bunsen burner for different times this would be an additional variable. Can you think of another method of getting the food to burn?
3. Do substances burn better in oxygen than air?
4. What about the volume of water? Should the water be stirred to distribute the heat?
5. Will all the heat go into the water in a boiling tube? If not, where will it go? Could you reduce heat losses? Why could this be described as a **systematic error**?
6. The boiling tube is made of glass. Is glass the best material to use or could you suggest a better conductor of heat?
7. How would you measure the temperature rise of the water? How would you make sure your results are **accurate**? How could you make them **precise**?

Systematic errors affect all results in an experiment. The results are shifted away from the true value because of an error or fault in the apparatus.

Zero errors are a type of systematic error, caused by using a measuring instrument that does not read zero before taking the measurement.

Random errors cause readings to vary from the true value; they can be compensated for by taking a large number of readings. The **true value** is the value you would get if you could measure the quantity without any errors at all.

A **fair test** is one in which only the independent variable is allowed to affect the dependent variable. All other factors that could affect the outcome are kept constant.

Accuracy is how close results are to the true value.

Calibration involves dividing the scale on a measuring instrument (e.g. a thermometer) between the highest and lowest values it can measure.

Precision is related to the smallest scale division on the measuring instrument.

After you have evaluated this experiment, look at the diagram of the food calorimeter in Figure 2.8 and see how many of your criticisms have been overcome.

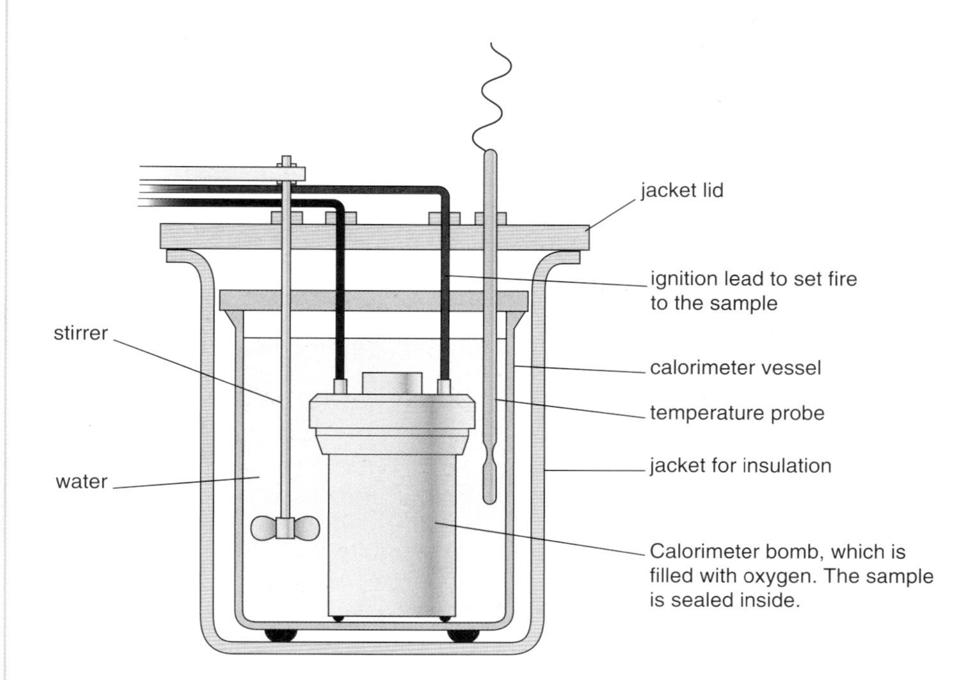

Figure 2.8 A food calorimeter. Experiments with food calorimeters show that 'one gram of fat contains more than twice as much energy as one gram of carbohydrate'.

Should we avoid eating all fats?

A high-fat diet can lead to obesity. Read the following facts and then decide if we should avoid eating all fats. Fats have important functions in the body:

- they protect vital organs from damage;
- they form an energy reserve;
- essential fatty acids (EFAs) are used in our bodies to make hormones, including the sex hormones;
- they contain cholesterol which is important for making strong cell membranes;
- fats contain fat-soluble vitamins (A, D, E).

Sorting out the fats

The terms saturated and unsaturated are used to describe different fats.

Saturated fats are hard at room temperature. They include fat around meat or left in the pan after cooking.

Unsaturated fats are more likely to be liquid at room temperature. The word 'unsaturated' is used to describe the double bonds between some carbon atoms in their structure.

Figure 2.9 Dairy products and red meats contain saturated fats. Saturated fats can be used for energy but excess is stored in the body as fat deposits. Too much saturated fat can raise your blood cholesterol level.

Figure 2.10 Olive oil, peanuts and avocados contain mono-unsaturated fats. These can be used for energy and help to lower blood cholesterol levels.

Figure 2.11 Oily fish, sunflower seeds and vegetable oils contain polyunsaturated fats. They also contain essential fatty acids called omega-6 and omega-3 which help to lower the blood cholesterol level. Even though the EFAs have essential roles in the body excess is still stored in the fatty tissues.

3. Nutritionists say that saturated fats should be limited to about 10% of the daily recommended energy intake for an active teenager. Why do they give this advice?
4. Nutritionists also suggest that we should eat oily fish, such as salmon or sardines, at least twice a week. Why do they give this advice?
5. Although there are more and more people who are obese, the amount of energy in the food that people eat has decreased slightly. Suggest a possible reason for these contradictory statements.

Look on the side of a food packet and you will find the 'nutritional information' which has a list of the fat types present.

Both mono- and polyunsaturated fats help to reduce cholesterol, but the polyunsaturated fats contain EFAs. Figures 2.9–2.11 will help you to picture these different fats.

Fats are sometimes combined in molecules with proteins. These molecules are called lipoproteins. Lipoproteins are important in our bodies because they transport cholesterol in the blood. High-density lipoproteins (HDLs) carry cholesterol out of the blood vessels and this is sometimes called 'good' cholesterol. Low-density lipoproteins (LDLs) are responsible for depositing cholesterol in blood vessels and so this is sometimes called 'bad' cholesterol.

Cholesterol is important in the body for making strong cell membranes and vital hormones. The cholesterol level in the blood depends on:

- the amount produced by the liver (an inherited factor);
- the amount eaten in food.

If the cholesterol level in the blood is too high, fat can be deposited in the lining of blood vessels. This raises blood pressure and increases the risk of heart disease. Women have a high level of HDLs until the menopause. The HDLs transport cholesterol out of blood vessels to the liver to be broken down. HDLs protect younger women from fatty deposits. Men have lower levels of HDLs and higher levels of LDLs so they are more at risk of blocked blood vessels, especially if they eat too much saturated fat and also smoke.

6. What sort of foods contain cholesterol?
7. Some diets are advertised as 'cholesterol free'. Is a cholesterol free diet a good idea? Give the reasons for your answer.
8. Earlier in this chapter we raised the question 'Should we avoid eating all fats?' Give at least five reasons why your answer should be 'No'.
9. Look carefully at the following statements. Which do you think is the best explanation of childhood obesity? Explain your choice.
   a) 'Children get fat because they watch too much television or play too many computer games.'
   b) 'Children get fat because they watch television rather than take part in active outdoor games.'
   c) 'Children get fat because they snack all the time while watching television.'

The nutritionist's viewpoint
Nutritionists have come to the conclusion that 'obesity is a complex problem caused by the interaction of many factors' and there is no single 'quick-fix' solution.

Using 'nutritional information' on food packets

Nutritionists recommend that the proportion of saturated fats in our diet should be about 10%. We can use the nutritional information on food packaging to check this.

A small 50 g bar of chocolate contains 30 g fat per 100 g chocolate.
So, the mass of fat in the chocolate bar = 15 g.
(If saturated fats are not listed separately a reasonable estimate is half of total fats.)
Therefore, the mass of saturated fats in the chocolate bar = 7.5 g.

1 g fat provides 37 kJ of energy.
Therefore, the saturated fats in the chocolate bar provide
$7.5 \times 37 = 277.5$ kJ of energy.

The average female teenager requires 8100 kJ of energy per day and 10% of the energy should be saturated fats – this is 810 kJ.

So, the small chocolate bar provides $277.5 \div 810 \times 100\% = 34\%$ of the daily recommended total of saturated fats.

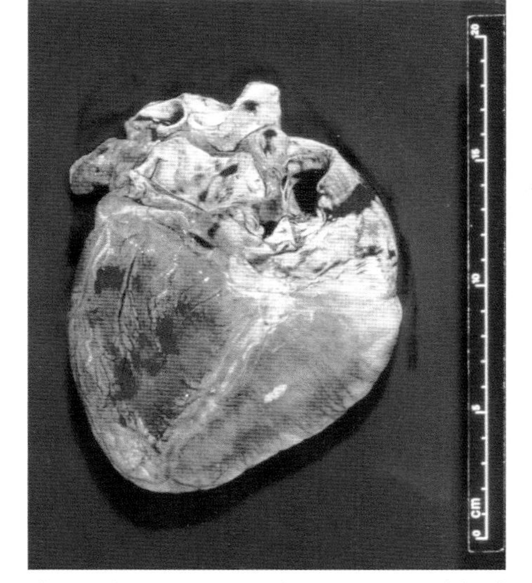

Figure 2.12 Save your heart! Don't add salt to food at the table or in cooking. Be more adventurous and add herbs for flavouring.

Salt in our diet

Salt is added to food as a preservative and to 'bring out the flavour'. It is very easy to exceed the recommended salt intake. Children in particular are at risk of consuming too much salt if they eat ready meals and crisps. There is very little salt in raw or fresh foods. The salt intake for a 7–14 year old should not exceed 5 g per day and for over 14s and adults it should not exceed 6 g per day.

Doctors are concerned about the amount of salt that people have in their diet. High levels of salt are linked to high blood pressure, which is a risk factor for heart disease and strokes. Heart disease is a major cause of death in the UK.

Activity – What is the proportion of fat in your diet?

a) Keep a food diary for one day. Write down exactly what you have eaten after each meal or snack because it is easy to forget.

b) Have a look at the 'nutritional information' on the packaging of the foods you eat. Alternatively you could check them on the internet. Find the amount or type of fat contained in your food.
Construct a table using the following headings. You will need one row for each different food eaten.

Mass of food eaten (g)	Mass of fat (g)	Mass of saturated fat (g)

Table 2.2 My food diary

Use the information in the section above to complete the following activities.

c) Complete your food diary and calculate your total mass of fats and saturated fats.

d) Does your saturated fat intake exceed 10% of your recommended energy intake? (Remember that the recommended energy intake for a female teenager is 8100 kJ per day and for a male teenager is 10 000 kJ per day.)

e) Which food gives you most saturated fat?

f) What problems might you have if your total fat intake is too low?

g) What foods should you eat only in limited amounts if you want to 'eat for health'?

h) Suggest five ways to reduce your fat intake.

How do our energy and nutrient requirements change with age?

Figure 2.13 Pre-school children are small but growing and they are very active. They need a varied diet which provides:
- enough energy for their activity;
- a good supply of protein for growth;
- calcium for bone formation;
- iron for haemoglobin in the blood;
- the fat-soluble vitamins A and D for general good heath.

The minerals and fibre should come from fruit and vegetables. They should eat small regular meals because of their small stomach size.

Figure 2.14 Pre-teens should:
- eat food from all categories listed in Table 2.1;
- eat more starchy foods than pre-school children, such as wholegrain cereals and potatoes for energy;
- avoid too much fatty food such as cakes, biscuits, chocolate and ice-cream.

Figure 2.15 Teenagers should be aware of what they are eating. The teenage growth spurt should be supported by:
- an increase in protein foods;
- an increase in iron for male muscle building and female menstruation (periods);
- sufficient calcium-rich foods for bone development;
- 'five-a-day fruit and vegetables' to provide minerals and vitamins for general good health and clear skin;
- drinking milk (semi-skimmed rather than full-fat milk to reduce fat intake).

Figure 2.16 **Pregnant and breast-feeding mothers** are eating for themselves and a baby (not for two adults). They therefore need:
- more calcium for bones and teeth;
- a little more protein;
- fruit and vegetable to give them plenty of vitamins and minerals.

Figure 2.17 **Adult** diets should be related to exercise. For example, a window cleaner will need to eat more than an office worker. Adults need to eat a balanced diet including all food groups but avoiding excess fat, sugar and alcohol.

Figure 2.18 **Older people over 65** tend to reduce the variety of their food and suffer from vitamin and mineral shortages. They should:
- continue to eat a range of fruit and vegetables;
- eat some protein every day;
- change their energy intake to balance the exercise they take.

2.2 Exercise for health

Activity	Energy used in joules
Walking slowly	314
Walking quickly	628
Pushing electric/petrol lawnmower	700
Cycling at normal speed	750
Aerobics	810
Swimming (steady crawl)	820
Tennis doubles	630
Tennis singles	1000
Running 1 km in 6.2 minutes	1250
Running 1 km in 4.7 minutes	1700

Table 2.3 The energy requirements of different activities
The table shows the energy a person weighing 60 kg (approx 9.5 stone) would use in 30 minutes of different activities.

More energy is used if:
- your weight is greater;
- you are fit with well-developed muscles.

10 a) Which foods should a pregnant woman increase in her diet?
b) What is the reason for each increase?

11 a) If you eat several bars of chocolate and bags of chips what might you be taking in excess?
b) What health problems might result if this habit continued for a long time?

12 Heart disease is a problem in the UK. What could you do to reduce your risk of heart disease?

13 Haemoglobin in red blood cells transports oxygen. Which mineral is needed for the production of haemoglobin?

14 Look at the figures in Table 2.3. How does energy use change with speed?

15 a) Copy the table below and calculate the total energy used by Helen in a typical day. Assume her body mass is 60 kg.

Activity	Time in min	(Show your working here)	Energy used in kJ
Helen cycles to school and home again taking 7.5 minutes each way	15	Energy used in 30 minutes of cycling is 750 kJ Energy used in cycling for 15 minutes = 750 ÷ 2	375
On arrival she does one hour of swimming training			
At lunchtime she walks around slowly while chatting	15		
In the evening she plays tennis (doubles) for an hour			
		Total	

Table 2.4 Helen's activity record

b) Copy the table below and calculate the total energy used by Saroj in a typical day. Assume that his body mass is 60 kg.

Activity	Time in min	(Show your working here)	Energy used in kJ
The day starts with Saroj's paper round. This takes him 30 minutes walking quickly			
He is late so he runs the 2.0 km to school getting there in 12 minutes			
He rests at lunchtime		*0*	*0*
He walks home slowly, taking an hour to get there			
In the evening, he earns more pocket money by mowing an elderly person's lawn. He takes 45 minutes using an electric push mower			
		Total	

Table 2.5 Saroj's activity record

c) Now make a similar table to calculate your own energy use on a typical school day. How could you increase your exercise level?

Metabolic rate is the rate at which food is broken down chemically in our bodies. It can be measured in joules per minute.
Muscle tissue has a higher metabolic rate than fatty tissue so a muscular person uses up more energy than a skinny or fat person.

Environment or genetics?

Our diets, the exercise we take and the **metabolic rates** of our bodies are affected by genetic and environmental (lifestyle) factors. Humans evolved as active beings who were able to survive in an environment where food was sometimes in short supply. Our ancestors were hunter-gatherers and farmers. They had a **good energy in / energy used balance**. Today's car trip to the supermarket rarely finds any food shortages and does not use up much of our energy intake. In many parts of the world our lifestyle has changed significantly over the years.

- We use a car or bus to get around, rather than walk.
- Advertising companies urge us to eat high-energy foods such as crisps and canned drinks.
- Many people have desk-based jobs.
- We spend hours playing computer games, watching TV and DVDs – 'couch potatoes' and 'mouse potatoes' don't use much energy.

So compared with our ancestors we have a double disadvantage – **more energy (food) in and less energy used (exercise)** in our daily lives. The consequence of this is gain in weight.

How can we achieve the *energy in* = *energy used* balance to prevent weight gain?

In order to balance *energy in with energy used* and prevent weight gain we should:

- take regular exercise;
- be aware that cooking methods (such as frying) increase the energy in foods;
- know which foods contain hidden fats and sugars that provide surplus energy.

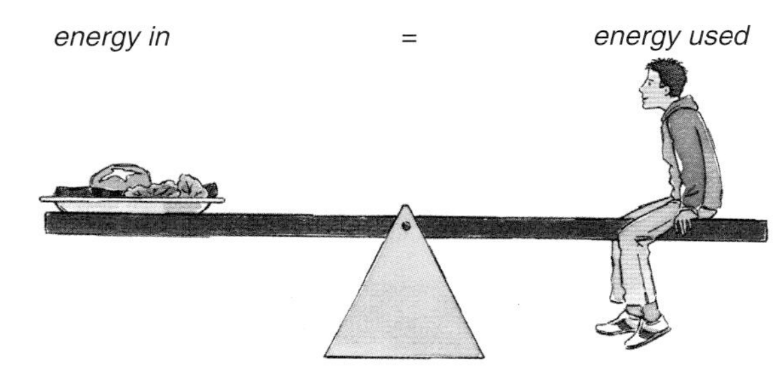

Figure 2.19 Balancing our food energy intake with the energy we use in our lives. Energy in = energy used

The benefits of exercise

Most gyms record a great increase in members around January 1st! Many people start with lots of enthusiasm but then stop attending as regularly as they did. Why do you think this is?

Ten good reasons for exercise are given on page 28 but it is important to remember that exercise needs to be both regular and enjoyable. You won't develop a 'six-pack' overnight, but firm well-toned muscles are well worth the effort.

Exercise is beneficial because it:

- increases the strength of your heart muscles and reduces blood pressure;
- improves lung ventilation (more air moved in and out of the lungs);
- improves oxygen uptake into the blood, which benefits your brain and body;
- increases muscle size and efficiency;
- strengthens ligaments and tendons and reduces the risk of injury;
- increases bone density and reduces the risk of osteoporosis (bone wastage). This disease affects both men and women but is more common in women after menopause. It can lead to fragile bones that break easily;
- decreases cholesterol levels in the blood and reduces the risk of blocked blood vessels;
- improves resistance to infections;
- raises metabolic rate and continues 'to burn up food' after you have finished exercising;
- leaves you with a 'feel good factor'.

If you check back, you can see that exercise and diet work together on improving the health of the whole body.

Activity – Food, health and exercise

1. Working as a group, check through a typical day and list:
   a) the high-energy foods you ate or avoided;
   b) the activity opportunities you took or avoided.
2. Is your life healthy or could you still make some positive changes for a healthier lifestyle? Remember, any changes must be fun if you are going to keep them up.
3. Doctors say that little changes such as walking up stairs, taking a slightly longer route to walk home and using a bike instead of getting a lift are more beneficial than big ideas such as 'I'm going to jog five miles three times a week.' Can you suggest some reasons why?

What happens when the energy balance is upset?

A big problem
Obesity is a global health issue with 310 million people affected.

Energy in food eaten is greater than energy used. The calculations in question 15 on page 26 show a direct link between exercise and energy used. Just look back at your conclusion. The key facts are:

i) the less exercise you take, the less food you need;
ii) if you take in more energy than you use, the excess is stored as fat.

In recent years, a lot of publicity has been given to obesity and the health problems it causes. In the past these were considered problems of 'old age' but now the same problems have been found in overweight teenagers and young adults.

Someone is regarded as overweight if they have a body mass index (BMI) greater than 25 and obese if they have a BMI of 30 or more.

Body mass index (BMI) is defined as $\frac{\text{body mass (kg)}}{\text{height}^2\ (\text{m}^2)}$

Body mass index (BMI) gives only a rough guide because it applies to over 18 year olds.

BMI does not take account of:
- the fact that muscle is denser than fat so athletes and body builders may have a BMI above 25 and still be healthy;
- variations in bone mass.

What are the problems of being overweight?

Overweight people are much more likely to suffer from arthritis, diabetes, high blood pressure and heart disease. These were once considered diseases of the 'over 50s'. Since the 1990s rates of these illnesses have been increasing. Some diseases, such as diabetes and rising blood pressure, are even seen in teenagers and young adults who are overweight. Children today consume the same food energy as children did 10 years ago, but they take far less exercise and so their energy input and energy use are not balanced.

Figure 2.20 'I don't enjoy football, I can't run as fast as the others'

The number of children with diabetes is increasing. Most of these diabetics take insufficient exercise (exercise uses up blood sugar) and 95% of them are overweight. (Section 8.1 contains more details on diabetes.)

With such a high percentage, it is clear that there is a link between:
- diabetes and being overweight;
- diabetes and not doing enough exercise.

It is not surprising then that the treatment of diabetes in young people is to stick to a planned diet and do lots of exercise.

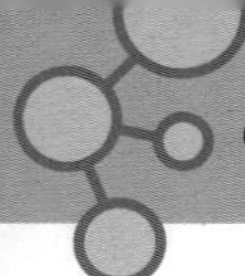

Diabetes in young people is taken very seriously because of increased blood pressure, which results in heart disease and damage to nerves, the retina of the eye and the kidneys.

Being overweight also puts extra pressure on the joints, particularly those at the knee. This results in worn joints or arthritis, which can become a painful problem well before middle age.

What are the problems of being underweight?

If the *energy we take in with our food is less than the energy we need for our lifestyle*, this can also cause problems.

You have probably heard of anorexia. This is a condition in which someone controls their own food intake so that it is well below that needed for healthy living. Anorexia affects about 1% of females. It is the third most common long term illness in teenagers, with a high mortality rate.

People also become underweight when there are food shortages. This often occurs when drought, famine or flooding hit developing countries. In these circumstances, people suffer from:

- deficiency diseases as a result of shortages of vitamins, minerals and essential amino and fatty acids (see Section 2.1);
- malnutrition as they are short of protein. Children are affected more than adults and have swollen bellies caused by water retention;
- a low resistance to infection as they are unable to form antibodies in response to pathogens (see Section 2.6). Epidemics such as measles spread rapidly through refugee camps;
- protein and iron shortages (in women). Their bodies respond by stopping periods, which helps them retain iron and protein and they can't become pregnant;
- mothers struggle to breast-feed their babies and so feed them on low-energy starchy foods with very little protein.

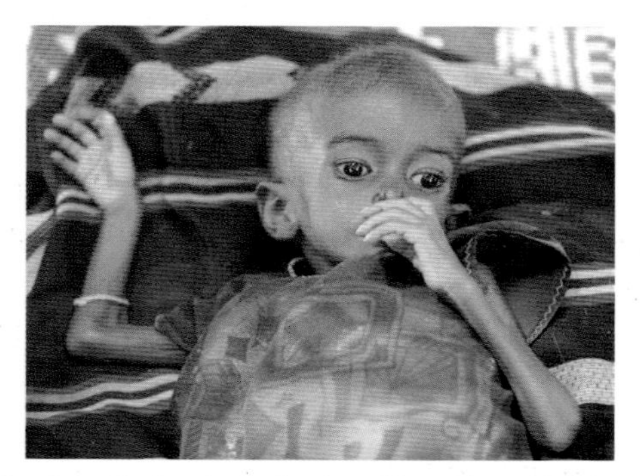

Figure 2.21 This toddler is suffering from malnutrition. The child has muscle wastage, no fat beneath the skin and is weak and lethargic.

The problems are made worse if the water supplies are contaminated with sewage. Water-borne diseases, such as cholera, are transmitted by drinking or cooking with contaminated water. Cholera causes diarrhoea and results in dehydration which is dangerous for babies.

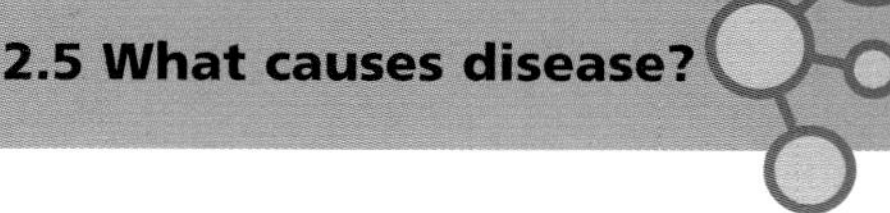

Slimming

People who are slimming try to **reduce their energy intake so that it is less than the energy used**. Their aim is to lose weight and have an attractive well-toned figure. The best way to achieve this is to eat wisely and to exercise more. Although slimming is now an outdated term, it still represents a multi-million pound business. Slimming diets have now been replaced by healthy eating and fitness plans.

Any diet should be examined to ensure that it provides a wide variety of foods including all the nutrients required for good mental and physical health. All the food groups in Table 2.1 should be included, particularly fruit and vegetables. Any diet based on a single food, such as bananas or eggs, is not healthy in the long term. The aim must be to achieve a healthy eating and exercise plan so that weight remains constant with a body mass index (BMI) of between 20 and 24.

Activity – Evaluating slimming diets

In groups:

a) Collect three different slimming diets from magazines or books.
b) Evaluate these diets carefully using the following questions for guidance. You could produce a tick chart in response to each of the questions and then add up the ticks at the end.
c) What is the best way of presenting your findings to the rest of the class?
   - Will all the food groups listed in Table 2.1 be eaten each day?
   - Is there the correct proportion of saturated fats?
   - Will the essential fatty acids be eaten?
   - Will the diet provide five servings of fruit or vegetables each day?
   - Are the foods colourful and 'fun' to eat or are they boring and the same each day?
   - Will the foods be satisfying and prevent feelings of hunger?
   - Is the quantity of salt less than 6 g per day?
   - Is exercise included?

In the last four sections we have seen that what we eat and how active we are can affect our health. In the next three we look at the topic of disease.

What causes disease?

Diseases are caused by **pathogens**. Pathogens are microorganisms which include bacteria, viruses and fungi. They are called microorganisms because they are incredibly small and can only be seen with a microscope. Different pathogens invade different parts of the body. This helps doctors to identify which disease we have caught.

The smallest microorganisms are viruses. Viruses:

- are taken into cells of the body;
- use the cell contents to **replicate** and form thousands of identical copies;
- damage the cell as they burst out;
- infect nearby cells and repeat the process.

Pathogens are microorganisms (bacteria or viruses) which cause disease.

Replicate means 'form an exact copy'. As viruses are not able to live independently they are not classed as 'living' so we do not use the word reproduce. They replicate using the DNA of the host cell. The damage to the host cell causes it to die.

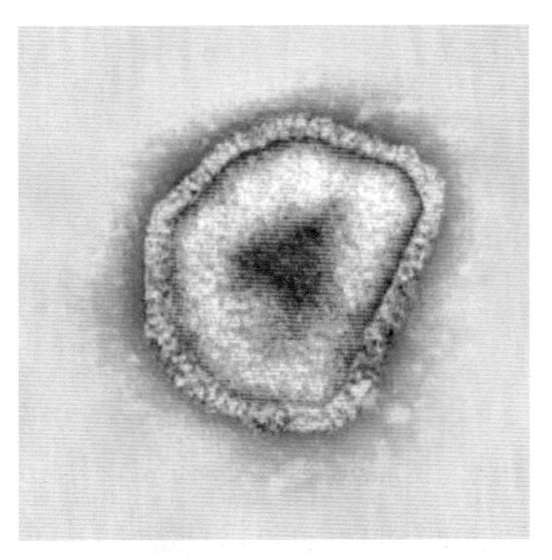

Figure 2.22 Photograph of virus taken using an electron microscope

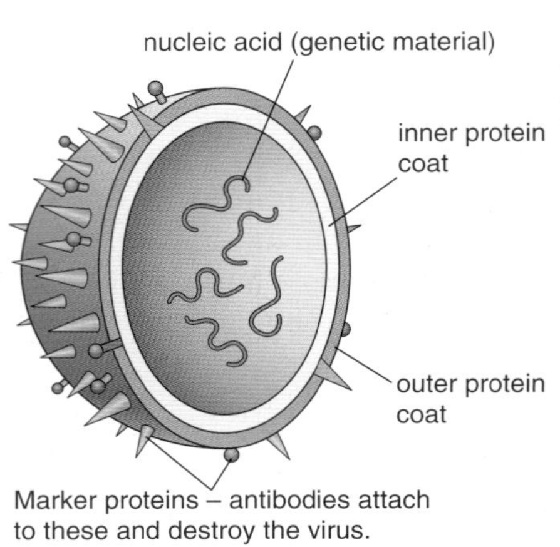

Figure 2.23 The structure of a virus

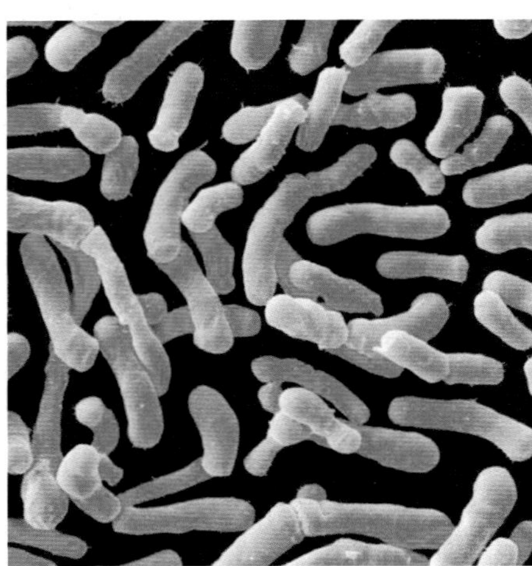

Figure 2.24 A photograph of bacteria taken with an electron microscope

Bacteria are much larger than viruses but can only be seen using a microscope. Pathogenic bacteria live in parts of the body such as the nose and throat but not inside cells. Some bacteria produce waste products which act as toxins and irritate cell membranes. These can cause sore throats and runny noses.

Figure 2.25 Ignaz Philipp Semmelweiss – an observant doctor

16 What do you think caused the difference in the death rate in the two wards?

17 Which are larger, bacteria or viruses?

Hygiene

We learn to wash our hands before eating or cleaning a cut when we are very young. But, it hasn't always been so. In the 1800s many women died of blood poisoning following childbirth.

Ignaz Philipp Semmelweiss, who was a doctor at a poor-mothers Maternity Hospital in Vienna, made the following observations:

- 29% of mothers died in the ward which doctors ran compared with only 3% of mothers in the ward run by midwives;
- few mothers died when giving birth at home;
- admissions after birth led to fewer deaths;
- only doctors carried out post-mortems (investigations to find the cause of death);
- doctors did not wash their hands between patients or after post-mortems;
- a doctor who cut his hand at a post-mortem died of the same blood poisoning as his patients.

Semmelweiss ordered all doctors to wash their hands between patients. They objected at first, but then saw the death rate drop to 1% and realised he was right. In 1861 Semmelweiss published his findings which were supported by the data he had collected. Many of the older doctors and scientists ridiculed his ideas and he died in 1865 before his simple lifesaving procedure became an accepted practice. (See MRSA in Section 2.7.)

2.6 The body's natural defence against pathogens

Antibodies are proteins produced by white blood cells that destroy pathogens which cause disease.

18. How does an antibody 'fit' onto a pathogen?
19. After the antibodies have caused the pathogens to stick together in a cluster, how are they destroyed?
20. If a person develops an infection why does it take a few days before they start to get better?
21. Why is a small child unlikely to catch chickenpox a second time?

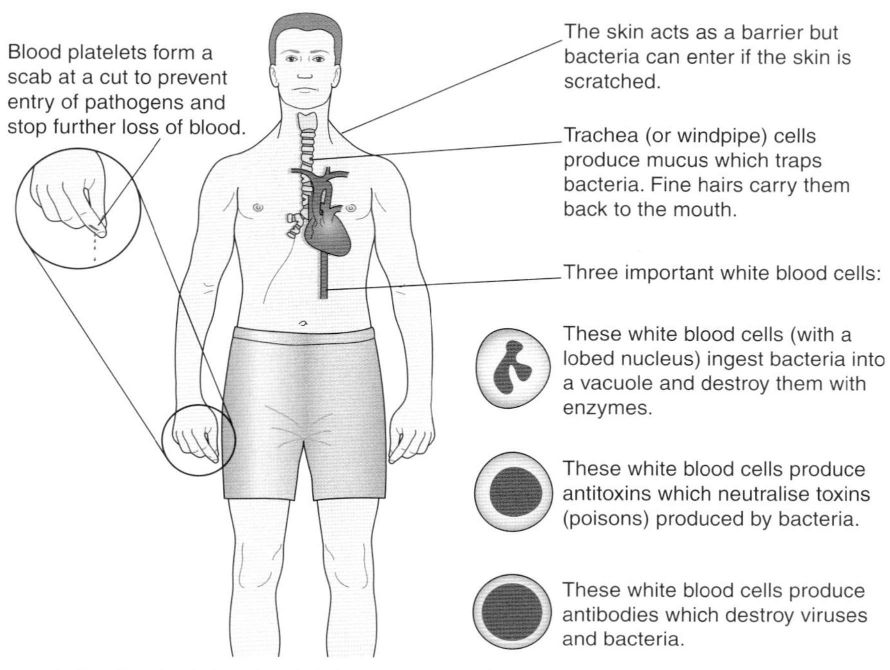

Figure 2.26 The body's natural defence against disease

All pathogens have unique marker proteins on their surface. These unique marker proteins are called antigens.

When someone catches an infection, their white blood cells respond by producing **antibodies** specific to these antigens. (Specific means that the antibody has a structure which only fits one antigen so there are as many different shaped antibodies as there are different antigens.)

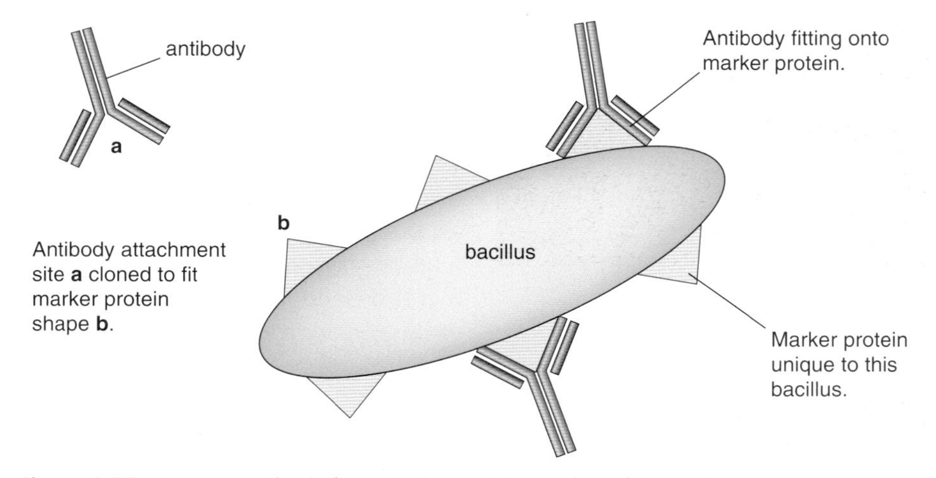

Figure 2.27 How an antibody fits onto the protein marker of the pathogen

Antibody production takes a few days. The antibodies cause the pathogens to stick together and these groups of pathogens are then ingested by white cells, as in Figure 2.26. Once the antibodies have been produced, the patient begins to get better. 'Memory cells' are also produced at this point. These 'memory cells' produce antibodies rapidly if the same pathogen invades the body again. This means that a person will not catch the same infection twice.

Protecting our bodies against pathogens

A **vaccine** is a solution which contains weak, inactive or dead pathogens.

A **vaccination programme** is a schedule of injections given to a child to prevent common childhood illnesses.

Our bodies can be protected against some pathogens by immunisation. This involves injecting a **vaccine** which contains weak, inactive or dead forms of a pathogen. The vaccine stimulates the white blood cells to produce antibodies specific to the pathogen. 'Memory cells' are also made which make the person immune to further infection by the same pathogen.

Vaccination can also be used to protect children against pathogenic viruses such as measles, mumps and rubella (MMR). When viral vaccines are prepared, the viral genetic material is removed and only the viral coat is used in the vaccine (see Figure 2.23). Following vaccination, antibodies and 'memory cells' are produced. A **vaccination programme** gives the child lasting protection (page 36).

Antibiotics

Antibiotics are medicines which kill bacteria inside the body which would otherwise cause further illness.

Antibiotics help to cure bacterial infections by killing bacteria inside the body. Different antibiotics attack bacteria in different ways. The most common antibiotic is penicillin. This makes the bacterial cell wall porous so that the bacteria burst. Other antibiotics prevent the bacteria from reproducing and some interfere with bacterial enzymes.

MRSA is an example of a strain of a common bacterium which has undergone a mutation. The mutation has changed the bacterial chemistry so the MRSA strain is resistant to most antibiotics. Overuse of antibiotics and natural selection of the resistant bacteria has resulted in MRSA becoming more common. Figure 2.29 explains this increase in resistant bacteria. If a patient with an infection is given an antibiotic, the normal bacteria are killed but any mutant bacteria may survive and divide, producing a large number of antibiotic-resistant bacteria. These can be passed on to other people. The best defence against MRSA is careful hand-washing. This has led to a major campaign in all hospitals with posters reminding staff and visitors to wash their hands or use alcohol gel before entering wards or touching patients. Hospital hygiene has improved but this is an area which needs constant attention to detail as bacteria are easily spread. Patients who become infected with MRSA are either isolated or barrier nursed. This means that the doctor or nurse wears a disposable apron and gloves whenever they touch the patient and throws these away afterwards. They must also wash their hands and use alcohol gel before visiting the next patient. This procedure has reduced hospital MRSA infections considerably.

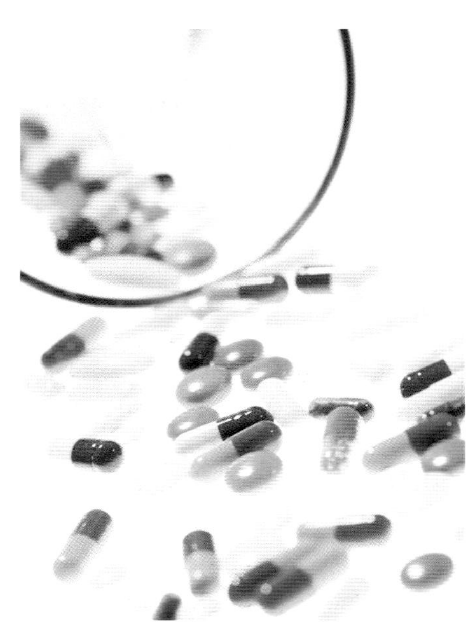

Figure 2.28 There are many different antibiotics

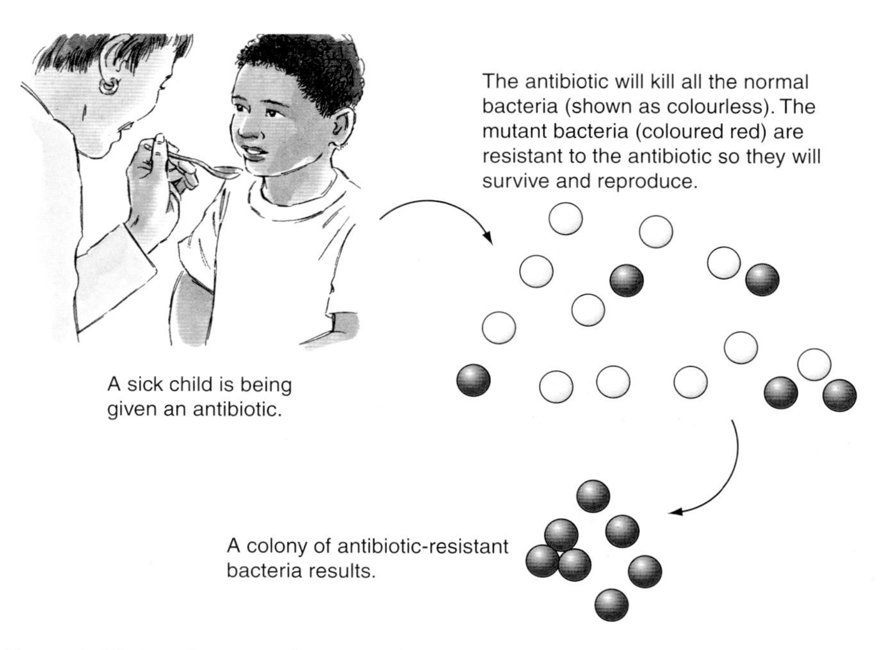

Figure 2.29 Development of a strain of bacteria resistant to antibiotics

Viruses

Unlike bacteria, viruses cannot be destroyed using antibiotics. Viruses live and replicate inside cells. It is therefore difficult to develop drugs which can enter cells and stop the virus replicating without damaging our body cells. We have to rely on our immune system to fight the virus.

Many people are worried about 'bird flu' (or more correctly, avian influenza). As you can see in Figure 2.23, viruses have a very simple structure. If the DNA from a bird flu virus were to get inside a human flu virus, the new mutated virus could spread through the human population very quickly. This is what happened in the mild flu epidemics of 1957 and 1968. The serious flu epidemic in 1918 was a bird flu virus that adapted to humans. People who rear chickens for meat and eggs have close contact with the birds and are therefore more likely to come into contact with the virus.

The virus is mostly in Asia but any cases in birds in other countries are closely monitored. Avian influenza could spread around the world as ducks and geese migrate to winter feeding grounds. Thousands of wild geese migrate to wetlands in Britain and other parts of Europe every year. The virus is excreted in the droppings of infected birds, so if a few carry the virus it could pass to domestic geese, ducks or chickens. The route from domestic poultry to humans is possible although the evidence is that the virus does not spread easily to humans from birds. However, AIDS is an example of a disease which has probably spread to humans from animals. Any human cases are monitored very carefully and so far no human to human transmission has been confirmed.

The HERD effect

In trying to prevent the spread of a disease in the younger population, the aim is to vaccinate 95% of all children in the UK. If this is achieved the pathogen cannot be transmitted because there are so few unprotected children. Vaccinating most of the 'herd' protects the others.

22 The vaccination programme explained in the bullet points on the right has been successful in preventing the spread of infectious disease in the UK. The conditions in a developing country are very different.

Use the bullet points on the right to explain why it is difficult to prevent the spread of an infectious disease in a developing country, such as Uganda.

Vaccination programmes

In developed countries deaths from infectious diseases are rare as a result of the **vaccination programmes**. Vaccination has proved very successful in the UK for the following reasons:

- there is a schedule for vaccination of babies and children up to 15 years of age;
- the majority of children are vaccinated;
- parents receive computer printed reminders for booster injections;
- there are good supplies of safe vaccine paid for by the government;
- there are refrigerated lorries to transport the vaccine;
- there are sufficient local community nurses to carry out the programme;
- most children are healthy and well fed so their immune systems respond by producing antibodies and memory cells.

Summary

- ✓ **A balanced diet** should contain different food groups. It must also provide essential amino acids and fatty acids, vitamins, minerals and the correct amount of energy.
- ✓ A diet that lacks any of the above items leads to poor health.
- ✓ If you take in a larger amount of energy than you use, the excess is stored as fat.
- ✓ High levels of **cholesterol** in the blood increase the risk of heart disease.
- ✓ Cholesterol is transported in the blood by lipoproteins. HDLs are 'good' because they remove cholesterol from the blood. LDLs are 'bad' because they cause fatty deposits in blood vessels.
- ✓ **Metabolic rate** is the rate at which the body uses food and energy.
- ✓ Exercise raises the metabolic rate.
- ✓ People who exercise regularly are usually fitter than people who take little exercise.
- ✓ In developed countries, arthritis, diabetes, heart disease and high blood pressure are linked to obesity.
- ✓ In developing countries, most diseases are related to food shortage and contaminated water.
- ✓ Too much salt in your diet is a risk factor for heart disease.
- ✓ Processed ready meals often contain high salt levels.
- ✓ **Pathogens** are microorganisms which cause disease.
- ✓ The body's protection against disease includes the skin, mucous membranes and white blood cells.
- ✓ **Antibiotics** are medicines which kill pathogenic bacteria in the body.
- ✓ **Bacteria** reproduce rapidly and can undergo genetic mutations. MRSA is a mutant strain which is resistant to many antibiotics.
- ✓ Hospitals can control the spread of bacterial infections by strict hygiene control.
- ✓ In the nineteenth century, Semmelweiss observed that hand-washing reduced the spread of infection.
- ✓ **Vaccines** stimulate the white blood cells to produce antibodies specific to the infecting pathogen.
- ✓ **Vaccination programmes** have been very effective against infectious disease.

EXAMQUESTIONS

1. Which of the following are causes of obesity?
   A lack of exercise
   B drinking too much water
   C overeating
   D eating seven servings of green vegetables each day *(2 marks)*

2. Which of the following could result from a diet with a large amount of red meat?
   A excess of calcium
   B saturated fats in the blood
   C anaemia
   D a high level of cholesterol in the blood *(2 marks)*

In questions 3–5, only one of the alternative answers is correct.

3. People who exercise are generally fitter than those who don't because
   A they have a low metabolic rate
   B they have small muscles
   C their heart muscles are stronger
   D they have high blood pressure. *(1 mark)*

4. Antibodies prevent infection by
   A causing bacteria to stick together
   B sterilising the skin
   C destroying bacterial cell walls
   D producing sticky mucus. *(1 mark)*

5. This question is about MRSA. Which statement is *incorrect*?
   A MRSA is resistant to many antibiotics.
   B MRSA has developed resistance as a result of mutation.
   C MRSA can be stopped from spreading by good hygiene.
   D A person infected with MRSA cannot be cured. *(1 mark)*

6. A person eats some food contaminated with food poisoning bacteria.
   a) Which part of the body will these bacteria infect? *(1 mark)*
   b) The person is ill for two days and then begins to feel better. Draw a diagram of the white blood cells that produce antibodies. *(1 mark)*
   c) Draw a diagram to show how an antibody is specific to one pathogen. Label this diagram. *(3 marks)*
   d) How do memory cells prevent a person catching an infection a second time? *(2 marks)*

7. a) How should doctors and nurses try to reduce infections passing from one patient to another? *(2 marks)*
   b) Give two ways in which antibiotics can kill bacterial cells. *(2 marks)*
   c) Explain how a bacterium can become resistant to an antibiotic. *(3 marks)*

Chapter 3
What determines where organisms live?

At the end of this chapter you should:

- ✓ be able to identify special adaptive features of animals;
- ✓ appreciate how adaptations allow an animal to survive in hostile environments;
- ✓ recognise the adaptations of plants for different environments;
- ✓ understand competition among plants and animals;
- ✓ understand the consequences of an increasing human population on the depletion of raw materials;
- ✓ be aware of the reasons for an increasing area of land use by humans;
- ✓ be aware of an increasing production of waste and pollutants by humans;
- ✓ be able to evaluate scientific evidence and separate this from non-scientific opinion;
- ✓ appreciate the factors involved in decisions about sustainability.

Figure 3.1 Commercial logging in an Indonesian rainforest causes widespread damage to the ecosystem

What determines where a particular species lives?

At KS3 you investigated food chains, food webs and 'who ate whom'.

In this chapter we will look, in greater depth, at the adaptations of organisms to the specific environments in which they live and the factors which control population size.

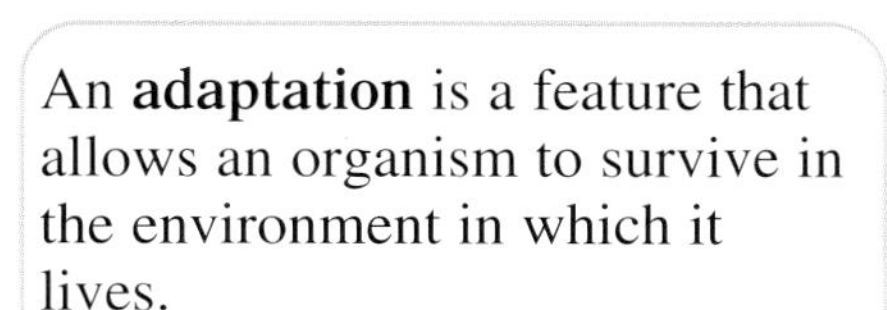

An **adaptation** is a feature that allows an organism to survive in the environment in which it lives.

Animals are said to be **adapted** to their surroundings when they are able to survive and reproduce successfully. In this case, the environment provides a supply of food for growth and a nest site or place to raise their young.

If you were going on a field trip to the Arctic, what clothing would you take? You would probably take a thick fleece and windproof jacket. Animals that live in the Arctic must also be able to withstand the cold conditions.

Figure 3.2 Polar bears are adapted to survive in cold conditions

The photograph in Figure 3.2 shows the features or adaptations which enable a polar bear to survive in the Arctic. The polar bear has:

- a small head and ears;
- a compact body shape which is also streamlined for swimming;
- a thick layer of fur – this traps air, which is a good insulator;
- a thick layer of fat (up to 11 cm thick) which insulates against heat loss and also acts as a food reserve during the Arctic winter when the bear is in a deep sleep;
- white fur which camouflages the bear, enabling it to get closer to prey when hunting;
- white fur to reduce heat radiated from the body.

Compare these features of the polar bear with a camel.

Figure 3.3 The Arabian camel is adapted to survive in hot, dry conditions

A camel lives in hot **arid** conditions. It needs to prevent overheating and is adapted to this by having:

- long legs and neck giving a large surface area for heat loss;
- thin hair on top of the body to allow heat loss;
- no hair on the underside of its body making heat loss easier;
- little body fat so heat is easily lost from the skin capillaries;
- a fatty hump which can act as a food reserve when desert food is scarce;
- camouflage colouring;
- two rows of eyelashes;
- nostrils which close for protection during sandstorms.

An **arid** region is hot and dry. This results in sparse vegetation which is usually adapted to reduce water loss and attack by herbivores.

Many plants are also adapted to live in harsh environments with features to stop them losing too much water. The house leek (Figure 3.4) lives on rocky outcrops where rainfall varies during the year with none at all during the summer. Similar conditions are found when they are grown on roofs or in rock gardens.

Figure 3.4 A house leek showing the fleshy leaves with a waxy coating

Water loss is reduced by having:

- a short stem;
- fleshy green leaves which store water but which dry up at the end of the year;
- a waxy, shiny outer covering to the leaves;
- long roots which penetrate deep into the soil in the rock crevices.

From these examples you can see that plants and animals can survive in different conditions because of the special adaptations which suit their environment. Look carefully at the photographs of two different foxes and answer the questions below.

Figure 3.5 Arctic fox

Figure 3.6 Fennec fox

1. a) Which animal is adapted to retain body heat?
   b) Explain how its body features help to retain heat.
2. a) How do the fur colours relate to their environment?
   b) The Arctic fox moults its coat in the spring. What colour would you expect the summer coat to be?
3. What is the purpose of the thick layer of fat on the Arctic fox?
4. The Arctic fox has short legs covered with thick fur. What would you expect the legs of the fennec fox to look like?
5. Why do the two foxes, which both listen for prey, have such different sized ears?

3.2 What determines the number of organisms in an environment?

Plants compete for light to photosynthesise and for soil space from which to absorb water and take in minerals. This interaction of organisms (plants or animals) trying to obtain the same food or occupy the same territory is called **competition**.

Consider a hedgerow beside a road. Plants such as primroses, like those in Figure 3.7, flower early in the year to avoid competition for light. They also produce leaves, flowers and seeds before the tree leaves open and put them in shade.

Figure 3.7 In this hedgerow the primroses have flowered before the trees have opened their leaf buds

Competition is the interaction between organisms trying to obtain the same food or occupy the same territory.

Animal populations are also regulated by competition among members of the same species and competition between different species. Some animals and birds compete for a territory of sufficient size to feed their offspring. Although people love to hear blackbirds and robins singing, their songs are actually 'war cries' to keep other males off their territory. These birds sing from different high points around the area to mark the boundary. Can you name any other animals that have a territory? How do they mark it?

Figure 3.8 A gannet colony. The nests are evenly spaced and just out of reach of the bird at the next nest.

Figure 3.9 Gannets have very pointed beaks.

Gannets are sea birds that catch fish by diving head-first into the water. They live and breed on remote rocks or cliffs. Just imagine the noise in the colony when all the birds are competing for mates and nesting sites.

Notice in Figure 3.8 that the nests are placed 'pecking distance' apart. You can see why when you look at the sharp, pointed beak of the gannet! The size of a gannet population changes with food availability. If more fish are available, more young gannets are raised, but this increases the competition for nest sites in future years. Other limiting factors within the gannet colony are **predators**, such as gulls, who brave the gannets' spear-like beaks to steal eggs or young.

A **predator** is an animal which catches and eats another animal. The animal caught is called the **prey**.

In general, the number of organisms in an environment is strongly influenced by food supply. Numbers will increase if there is plenty of food and decrease if food is in short supply. From the examples given, you can see that population size is also determined by competition among members of the same species, and by competition with other species for the same nutrients or space. This applies to both plants and animals.

3.3 How do humans affect the environment?

There are 6 billion people in the world today and the number is increasing.

How can the requirements of this increasing human population be met without damage to the environment? People in the UK expect to have a house or flat to live in and access by road to supermarkets and other

An **ecosystem** is made up of the plants and animals living and surviving in one place and interacting with the surrounding non-living environment.

A **biological control species** is an organism that limits the numbers of another species that is considered a pest. (For example, ladybirds eat greenfly – the ladybird is the control species and the greenfly is the pest.)

Figure 3.10 A road widening scheme carves through woodland, dividing the ecosystem

Activity – Campaign poster

A local council is proposing to build a new road through the middle of an area of ancient woodland which contains rare plant and butterfly species and a small population of deer. It is also a popular area for families at weekends. Prepare a campaign poster outlining the reasons for re-routing the road around the wood. Use illustrations to add interest.

shops. In their homes they want electrical goods such as a fridge, freezer and washing machine, and many have a computer and media system. Can all these be provided without impact on the environment? To answer this question we will examine the issues of a) land use; b) raw materials; c) waste, and d) waste-water treatment.

a) Does an increasing population reduce the amount of land available for animals and plants?

An increasing human population needs more land for:

- building homes, industrial estates, motorways and retail parks;
- quarrying and mining of raw materials for buildings and roads;
- intensive farming;
- waste disposal.

Currently between 10 and 20% of the UK's native species are threatened with extinction by expanding towns, motorways and intensive agriculture. Sites of Special Scientific Interest (SSSIs) have been established in some areas to support a rare species or group of species. Such species need special protection and the ecosystem around them needs to be large enough to remain balanced and undisturbed. Size is very important when setting up an SSSI. However, there can often be problems if a rare species is located where a motorway is planned!

A balanced **ecosystem** will change slowly over time. This means that the number of plants and animals and the range of species remain similar from year to year. Animals and plants need a certain area to sustain a breeding population. Dividing up an area of land, by putting a motorway through it, may result in the separate parts suddenly becoming too small to feed populations of larger animals such as deer or badgers.

Having examined the consequences of making an ecosystem smaller, consider the opposite case of making it larger. Removing hedges and ditches to make bigger fields removes the shelter, feeding and breeding sites for animals, including **biological control species**, such as predatory beetles, insectivorous birds and hedgehogs. These are the carnivores in the food web. By eating pests such as greenfly and caterpillars, these predators can naturally keep pest numbers low. Removing hedges and ditches upsets the balance of the ecosystem, disrupting the food webs and allowing pests to reproduce and cause further crop damage.

b) Does an increasing population have an impact on raw materials?

Important raw materials, such as building stone, metal ores and fossil fuels, are finite resources. Their increased use means they will run out more quickly and alternatives, such as renewable energy sources and recycled building materials, must be developed. Trees take many years to grow to a size where the wood can be used, so regular replanting must take place to prepare for future needs.

Figure 3.11 Landfill sites attract scavengers, such as gulls

6 Imagine an incinerator is being built near you and the campaign against its construction is being supported by your local newspaper.
a) Do you think the newspaper would report scientific evidence for and against the incinerator or people's feelings?
b) Do newspapers generally print scientific, factual information or eyecatching headlines?
c) Do you think a newspaper is always a reliable source of information, in such cases?
d) i) Would you expect to see a report with data in your local paper?
ii) If the data were produced by the company wanting to build the incinerator do you think they would be biased?
e) Technology and environmental regulations are regularly improved – do you think the newspaper information would be up-to-date?

c) Can the volume of waste deposited in landfill sites be reduced?

Low-lying areas of land and old quarry pits are often used as landfill sites, being filled with waste which is then compressed. Although these sites are well regulated there are still environmental problems associated with them. Sites can be unsightly. Pathogens can be transported by birds such as gulls, which feed on the waste, and toxic chemicals can be washed into waterways after heavy rainfall.

Landfill sites are not an ideal solution to the disposal of our rubbish. Many sites around the UK are now almost full and, in some cases, no local alternative site is available.

Waste disposal is a complex problem. New solutions to the increasing problem of waste disposal need to be found.

Currently:
- the UK produces over 100 million tonnes of waste per year;
- the increasing population of the UK and higher standards of living mean that waste production is expected to rise;
- a large percentage of rubbish is disposed of in landfill sites;
- a considerable proportion of the material placed in landfills could be recycled to save raw materials;
- the government initiative 'Waste Strategy 2000' plans to reduce landfill by 45% by 2010 and 67% by 2015.

So what can be done with all our rubbish? Alternatives to landfill include: recycling, incineration and composting. All of these alternatives require processing plants, which cost money to build although some of the costs could be met by the sale of materials.

Recycling

This can involve the mechanical sorting of mixed rubbish (for example, pulling iron / steel out using a giant electromagnet) or sorting by each household. See Section 3.6.

Incineration

This process further reduces the volume of rubbish placed in landfill sites. The ash produced can be used for road building, thus saving natural stone. All new incinerators are also EfW plants (Energy from Waste). This means that the heat produced from burning the rubbish is used to make electricity. Proposals for building new incinerators frequently encounter opposition from local residents who fear the risk of air pollution. Modern incinerators are, however, highly sophisticated industrial plants where all emissions are released at safe levels and monitored by a computer 24 hours per day.

Composting

This is used to reduce the volume of **biodegradable** materials. After treatment (to kill pathogens) the compost can be used or sold.

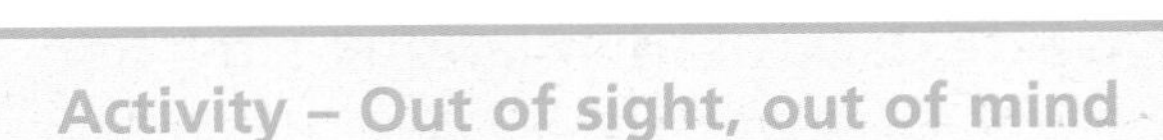

Activity – Out of sight, out of mind

It is easy to throw rubbish into a bin and forget about it.

a) Find out about each of the following waste disposal methods: recycling, incineration and composting. Construct a table of advantages and disadvantages for each.
You can use the internet to help you do this. Search for 'waste' on the 'UK Environmental Agency' website, www.environment-agency.gov.uk. Type in your postcode on the above website and find out how rubbish is dealt with in your area.
How is rubbish currently disposed of? Is there an obvious site such as the land of a disused factory or warehouse that could be used for a new treatment plant? Would rainwater drain from the site into a local river causing a pollution problem? Is the only access road for transporting waste to the site through the town centre?

b) Using your answers from question 6, write a letter to your local newspaper giving your opinion of which waste disposal methods should be used in your area. Give reasons for your decisions.

d) What is the impact of an increasing population on waste-water treatment?

Before the Industrial Revolution people lived in small, scattered rural communities. Waste water was emptied into ditches and the biodegradable contents decomposed. Now, towns have much larger populations. If sewage were emptied into our rivers it would create an awful smell and fish and other wildlife would be poisoned. The treatment of waste water and sewage has improved to meet the demands of an increasing population. Today only treated sewage passes into rivers or out to sea and it must not contain pathogens or toxins (poisons) or cause **eutrophication**.

Many pharmaceutical companies and power stations employ a biological solution to overcome waste-water problems. They filter waste water through reed beds. The reeds use the nutrients in the waste material to grow and at the same time remove toxins from the water. Reed beds provide an interesting habitat for wildlife. Water run-off from motorways, and some airports such as Heathrow airport, is treated by flowing through such reed beds.

Section 3.3 started with the question 'How do humans affect the environment?' After examining four different issues, we can now make some positive suggestions. As a society, we must:

- avoid changing balanced ecosystems as this can have serious consequences and damaging long term effects;
- increase recycling and the re-use of materials such as metals, plastics, glass and paper. This will save raw materials;
- reuse **brown field sites** in town for new housing. This will save farmland and new roads will not be needed;
- use biological methods where possible to treat waste water and biodegradable waste, as the end products are harmless to the environment.

Biodegradable material is usually of plant or animal origin. It can be decomposed by bacteria or fungi. (Paper, cotton and wool waste are all biodegradable.)

Eutrophication is the nutrient enrichment of a river by the addition of nitrates, phosphates or ammonium compounds. As a result, algae reproduce and the water soon becomes green. Rooted water-plants grow rapidly and this slows down the flow of water. Algae have a short life cycle. Death of the algae is followed by an increase in aerobic decomposer bacteria which break them down. These aerobic bacteria use up oxygen and this deoxygenates the water.

Figure 3.12 Reed bed filtration – an opportunity for a new ecosystem

7. An increasing population leads to increased transport. List three ways in which this can affect ecosystems.

8. The organisms shown in Figure 3.13 eat pests which damage plants.

   a) What is the name given to this useful group of insects?

   b) How have numbers of the biocontrol species been reduced?

Figure 3.13

3.4 Are the problems associated with an increasing population restricted to developed countries?

The term **brown field site** is used to describe an area in town where old properties have been demolished. A **green field site** is farmland or green belt land surrounding a town where there have been no buildings before.

Large-scale deforestation of tropical rainforests throughout the world has been, and still is, widely publicised. The Amazon rainforest is a major source of global oxygen and a large number of pharmaceutical drugs have been developed from rainforest plants. We therefore must take steps to ensure the survival of this resource. It also shows that the problems associated with an increasing worldwide population are not restricted to developed countries.

The ecosystem of a rainforest is dependent on nutrient recycling. Dead leaves are decomposed by bacteria and fungi, and the minerals are released into the soil. 'Slash and burn' also releases minerals into the soil which can support the arable crops of a small population for a few years. As minerals in the soil are removed and not replaced, the crop yields fall and the village population moves, abandoning the clearing. The natural vegetation will regenerate from seeds in the surrounding forest, but this takes time. Very different effects result from commercial logging (compare Figure 3.14 with the photograph at the opening of this chapter).

Figure 3.14 A small tribal clearing in the rainforest

Deforestation provides:

- timber that is sold for money;
- timber for use as a building material;
- land for cattle ranching or growing crops to generate income;
- land to build roads which connect major cities.

9 Why do small clearings cause less damage than large-scale logging?

10 a) Who do you think provides the evidence for the re-growth of vegetation in small cleared areas?
b) What type of evidence would you need to be convinced that regeneration takes place? Briefly suggest a project to investigate this.
c) What is it about the soil in a rainforest that limits its value as agricultural land?
d) Why does soil erosion occur in deforested areas?
e) i) What did scientists from the Massachusetts Institute of Technology discover?
ii) Do you think their evidence can be reliably linked to flood damage reported in vast areas cleared of forest?
f) List what you see as the harmful long term consequences of extensive deforestation of rainforest. Will the effects be local, global or both?
g) Using your answer to question f), suggest a management strategy that would result in sustainable development of the rainforest. (See Section 3.6 for an explanation of sustainability.)

Large-scale logging results in:

- vast cleared areas;
- removal of the tree canopy, which breaks heavy rainfall;
- compaction of the soil and root damage by heavy machinery;
- poor soil drainage and no aeration to the roots of small plants due to heavy machinery;
- soil erosion as water from heavy rainfall flows across the land;
- flooding when soil washes into streams and rivers and causes a blockage.

Scientists from the Massachusetts Institute of Technology examined 75 years of rain-gauge records. The records showed a big increase in rainfall over large deforested areas. The scientists also examined current satellite data of cloud cover. Satellite pictures showed twice as much low-level rain-bearing cloud over large deforested areas, compared with forested areas. Small deforested areas do not give the temperature differences needed for the formation of clouds.

Other effects of commercial deforestation include:

- isolation of animal populations which will not cross large open areas where forest has been cleared;
- reduction in the sources of food for both human and animal populations.

Approximately 160 000 square kilometres of deforested land have already been abandoned because the thin, acid soil cannot support crops or cattle. These areas are slowly becoming deserts.

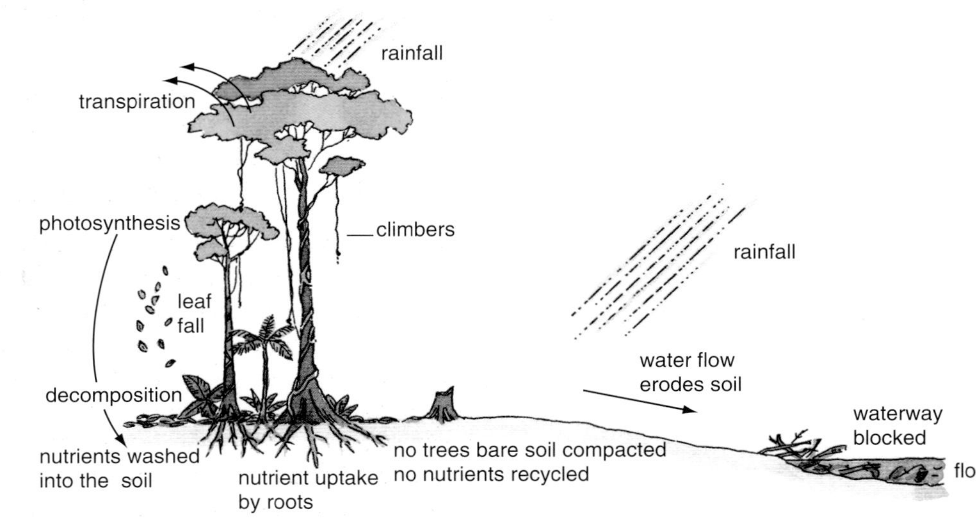

Figure 3.15 The rainforest cycle and biological consequences of disturbance

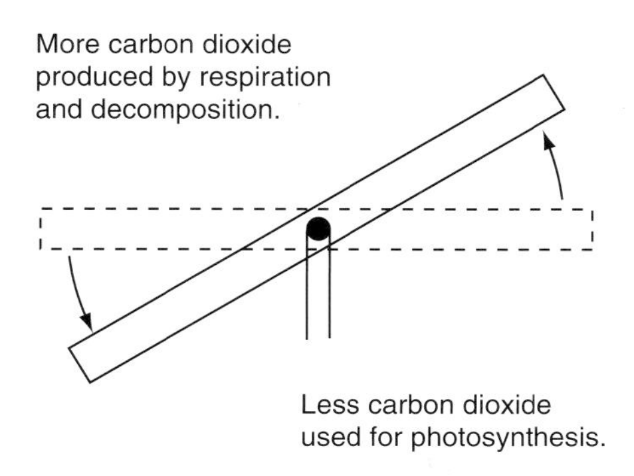

Figure 3.16 Deforestation results in an increase of carbon dioxide in the atmosphere

The Brazilian government uses satellite surveillance technology to help prevent illegal logging. It faces the dilemma of trying to improve the lives of the increasing Brazilian population while at the same time preserving the rainforest.

A rainforest appears lush but it is still a delicately balanced ecosystem (see Section 3.3). In an undisturbed ecosystem the carbon dioxide uptake for photosynthesis is balanced by that produced by respiration and decomposition. Logging results in the release of higher levels of carbon dioxide by:

- increased activity of soil microorganisms, as leaves and branches from felled trees decompose;
- burning of waste wood in fires.

Trees convert or 'lock-up' carbon from carbon dioxide into compounds such as cellulose during their period of growth. The ecosystem balance is upset when all the carbon that was 'locked-up' as cellulose is suddenly released in one fire. Slow plant regeneration means that there will be little photosynthesis as the ecosystem recovers and so the carbon dioxide remains in the atmosphere. Destruction of rainforest can therefore play a significant part in rising carbon dioxide levels.

Global warming

There is an interesting link between deforestation, cattle ranching and global warming. Deforestation provides land for cattle ranching. Cattle are ruminants. A large part of their stomach, called the rumen, acts as a fermentation chamber.

Anaerobic bacteria live where there is no oxygen. Methane (CH_4) is produced if carbon compounds are broken down when there is no oxygen present.

There are anaerobic bacteria inside the rumen that produce enzymes which break down cellulose and provide sugars for energy. One of the waste products of this process is methane, which is released into the atmosphere. Unfortunately methane is a potent greenhouse gas.

The anaerobic bacteria in the rumen of cattle are not the only methane producers. Methane is produced in paddy fields by anaerobic bacteria that decompose dead vegetation in the waterlogged conditions. As the population in Asia increases, more rice is required and so more land is flooded. This leads to an increase in the volume of methane released. Methane is also produced by the anaerobic decay of materials in sewage and household refuse. In some countries, this methane is collected, stored and used as fuel. In the UK the methane is burned and used to control the temperature of the anaerobic sewage digester tank.

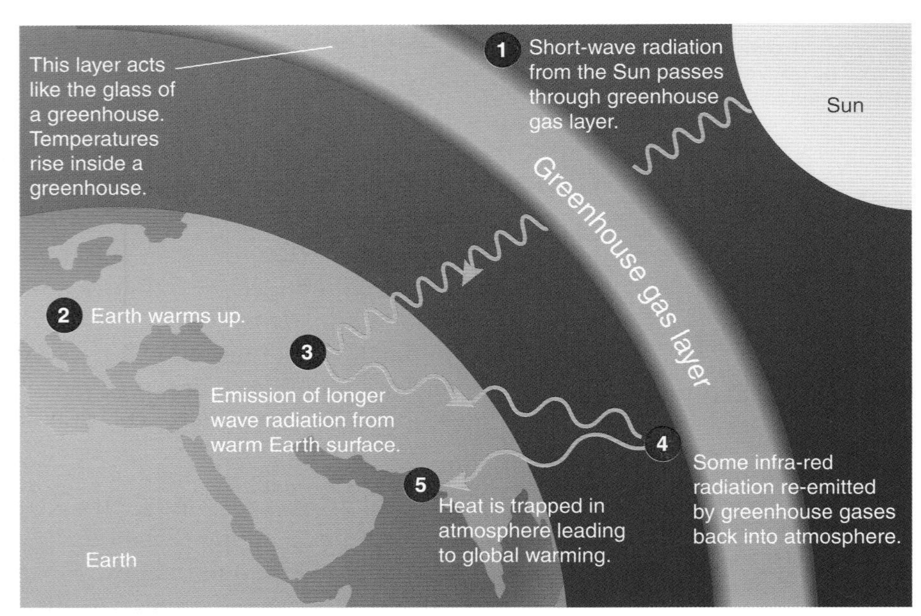

Figure 3.17 The greenhouse effect

A **scientific model** is a hypothesis which can be represented as a mathematical equation or a computer program. Models are valuable when there are many different factors which affect the outcome of an experiment. The accuracy of the model can be tested by applying it to results that have been collected in the past. If the model fits with past data, it can be used to simulate what might happen in the future. Weather forecasts are created using scientific models which incorporate many factors – if one factor changes unexpectedly the forecast may be wrong!

How science works – the use of models for complex problems

We have already seen that ever increasing amounts of carbon dioxide and methane are being produced. Most scientists believe that these so-called 'greenhouse gases' are largely responsible for global warming. Other scientists believe that global warming has taken place before and that we are in a new warming cycle which has nothing to do with these gases. When 'sceptics' challenge scientific ideas further research is carried out, further **scientific models** are produced and tested. With more detailed and reliable information scientists get closer to the truth. The 'global warming' debate takes place not only among scientists but also among politicians, environmentalists and humanitarians. Figure 3.18 shows the different bias which each group brings to the debate.

Figure 3.18 The global warming debate

Should we be concerned about the greenhouse effect and global warming?

By analysing records over many years, scientists have made the following observations:

- the cold winter period in the Arctic has become shorter and the ice caps are melting;
- sea levels have risen between 10 cm and 20 cm in the last hundred years as the snow cover in the Northern Hemisphere has melted;
- graphs of temperature and carbon dioxide show increases which follow similar trends;
- in recent years there has been unusually heavy rainfall and flooding in the UK and Northern Europe;
- droughts and crop failures in the Sudan and other regions have increased in recent years.

All these observations fit the predictions made by scientists and meteorologists (weather forecasters) based on their current scientific models.

Although there has been a great deal of debate about global warming, it now appears that scientists have underestimated the rate of rise in world (global) temperatures.

A rise in temperature of only a few degrees Celsius is enough to change the climate. This could mean a change in direction of rain-bearing winds, resulting in heavier rainfall in some parts of the world and severe drought elsewhere. If the melting of ice continues there will be a further rise in sea level and areas of Holland, eastern England and Bangladesh which are just above sea level could be flooded.

Having collected and analysed evidence from measured data and scientific models we can start to form conclusions. All governments should be concerned about global warming and they should take action to reduce their emissions of carbon dioxide and methane. Individuals should also play their part. This will be explored in the next section.

What can be done to reduce human impact on the environment?

Individuals, local government and industry can all make changes, based on scientific evidence, to reduce the negative human impact on the environment. People who recycle are acting responsibly to conserve natural resources. We *all* have a part to play in meeting the targets of local and national schemes to reduce waste. Local authorities are making recycling easier for us. Not only are there bottle banks and facilities for recycling aluminium cans, but many towns now also have kerbside collections for newspapers, bottles and cans. Countries such as Germany have been doing this longer than the UK and have significantly reduced waste.

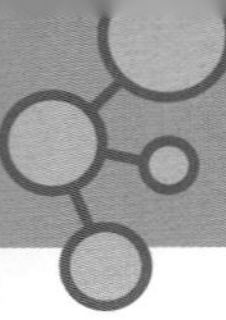

Figure 3.19 More and more young people are actively recycling and trying to reduce waste. Is there a point to this?

Activity – 'Waste local plan'

1. If your town has kerbside collections, find out which materials are collected separately.
2. Make a list of common products that can be produced from recycled materials.
   You may find www.recycledproducts.org.uk useful for your research.

The energy-efficient home: saving fuel saves fuel costs

Making our homes and offices more energy efficient by installing loft insulation, cavity wall insulation and double glazing reduces the amount of fuel needed for heating.

These changes are beneficial because:
- fossil fuels will last longer;
- greenhouse gas emissions will be reduced;
- gas, oil or electricity bills will be reduced.

11. Suggest other ways that you could reduce energy used a) in your home and b) within your school. Visit www.est.org.uk before answering this question.

Why walk or cycle?

Walking or cycling instead of driving:
- saves fuel costs of private vehicles;
- reduces air pollution;
- saves petrol and oil, which are non-renewable resources.

Public transport, which can carry many people, provides similar benefits to the environment.

Figure 3.20 Where is your nearest cycle route?

Activity – What can you do to make a better future?

This is a group activity. Log onto www.sustainable-development.gov.uk for some ideas.

Alternatively, you can log onto the RSPB website www.rspb.org.uk. Type 'green living' into the search box.

1. Planning
   In a small group, develop an environmental plan that involves avoiding waste or increasing recycling. Try to think of a new idea that you could put into action in either your school or the community in which you live. For example, a local charity might have a fund-raising recycling project for which you could provide support.
2. Communicating
   Create a poster, webpage or PowerPoint presentation which outlines your plan to others in your class. Remember to explain the importance of each suggestion in your plan. For example, recycling aluminium cans saves the raw material and the energy used to refine aluminium ore.
3. Action
   Obtain permission to carry out your plans and put them into action! Remember, you are more likely to keep up simple tasks, such as collecting cans for recycling at break or organising a petition for a can bank, than more complicated ideas.

Sustainable development is 'development to improve the quality of life for people living now without reducing the ability of future generations to meet their own needs' (United Nations definition).

Figure 3.21 The World Summit on Sustainable Development, Johannesburg 2002

Do governments consider the ethical issues when planning for future development?

Even within the UK there are issues of poverty, pollution and health care that need to be addressed. The UK government has already produced a **sustainable development** strategy. This means planning to live 'within the means of the environment' so that natural resources will still be available for future generations. One part of the strategy seeks to control fishing, to allow fish to grow to reproductive size and breed before they are caught. This should allow the UK fish population to increase and prevent fishing to extinction. Another part of the strategy encourages the redevelopment of brown field sites (previously developed land in towns) rather than developing and reducing the area of farmland or parkland, and building houses which ordinary people can afford.

You can log onto the UK's government website at www.sustainable-development.gov.uk for more information.

More details can be found at www.uyseg.org. Double click on the Sun to enter the site and then 'Chemical Industry Education Centre'. Click 'webcentre', 'secondary' and then 'sustain-ed'.

A brief history of the world efforts for sustainability

The first United Nations Conference, to consider the impact of human activity on the environment, took place in 1972. In 1992 more than 100 countries met for the first international Earth Summit. This summit produced a major plan for sustainable development, reducing poverty and giving people access to resources to support themselves. The plan called upon governments in all UN countries to lead the way in reducing pollution, emissions and the use of natural resources. In 2002 the Johannesburg Summit reviewed the little progress that had been made in the previous 10 years and focused on reducing poverty, and increasing access to safe drinking water and sanitation.

Activity – Easter Island

Figure 3.22 Easter Island was once a wooded island

Easter Island is known for its amazing statues, up to 10 metres high, which were carved in stone about 1200 years ago. Today, there are no trees on the island. But archaeological investigations have shown that the island was once heavily wooded when the first inhabitants arrived by canoe 1300 years ago. Easter Island palm trees were ideal for building houses and hollowing out for canoes which were used for catching seafood, especially porpoises. The statues provide evidence of a complex culture, because engineering skills must have been used in transporting the enormous statues to sites all around the island. This was done by rolling the statues on logs and pulling them with ropes made from climbing plants. The population on Easter Island increased rapidly but this resulted in deforestation followed by wind erosion of the soil, poor crop yields and the drying-up of water springs. Without palms to construct canoes, there were no porpoises to eat. Once the inhabitants of the island had eaten all the sea birds and shell fish, starvation, anarchy and cannibalism set in. This sad story was uncovered from the pollen record and bone deposits in rubbish sites on the island.

Is this a picture in miniature of our effect on other ecosystems? Year by year, there is evidence of similar damage to environments and tribes in the Brazilian rainforests, Africa and Asia. Is it time to act now for sustainability before it is too late?

1. Copy and complete the following table, comparing the disaster on Easter Island with tribal use and commercial logging of the Amazon rainforest.

	Easter Island	Tribal use of rainforest	Commercial logging of rainforest
Timber for	• Building homes • Building boats/ canoes • Rollers to transport statues • Fuel		
Main food supply		• Bush animals • Fruits, berries and roots • Grubs	Food is delivered to logging camps and results in food waste and packaging waste
Soil erosion	Yes – as a result of tree removal		
Animal populations	Reduced by over-fishing and habitat destruction	Little damage – small areas used in which plants regenerate quickly and animals spread back from nearby areas	
Problems faced by humans	Starvation as a result of over-fishing and habitat destruction		Conflict between displaced tribes and loggers
Can the environment be restored?			

Table 3.1

Living organisms as indicators of pollution

A **biological indicator** (**bioindicator**) is a plant or animal which can indicate the level of a particular chemical in its habitat. In the examples in this section sensitive lichens indicate sulfur dioxide levels in the air and invertebrates indicate oxygen levels in water.

Chemical analysis of water or air for pollutants gives a measure of what is present at the time of sampling, but it gives no information about pollution incidents which may already have been diluted or blown away. But, such incidents may kill certain species, leaving evidence for the biological detective.

Pollution indicator species

Different species of lichens have different sensitivities to sulfur dioxide (Table 3.2). Some are so sensitive that a trace of the gas in the air will kill them. This means that by looking at the lichens still growing, we can use them as **bioindicators** to find out the prevailing levels of sulfur dioxide pollution in the air.

Information from a lichen survey could help a local authority to monitor changes resulting from increased traffic density or industrial activity. This would enable it to make decisions about locations for new roads or industrial sites.

Lichens are pronounced 'li kens'. They are a mutual association of a fungus and an alga. They occur as crusty patches on rocks, tree trunks and walls. Lichens are living organisms which can be used as **biological indicators** (bioindicators) of pollution as the algal part is rapidly killed by sulfur dioxide in polluted air. Lichens grow very slowly over many years.

Type of lichen	1: Common orange lichen – quite tolerant of moderate pollution levels	2: Quite a common lichen but dies quickly if pollution levels rise	3: A common woodland lichen and an indicator of clean air	4: Beard lichen – only survives in pure air
Maximum level of sulfur dioxide tolerated	70 μg/m^3	60 μg/m^3	50 μg/m^3	35 μg/m^3

Table 3.2 Common lichens which act as bioindicators of air pollution from sulfur dioxide

Activity – Carrying out a pollution survey

A science class decided to survey the air quality around their town, using lichens as indicators.

Starting hypothesis: *The lichens growing will be those able to survive in the level of sulfur dioxide pollution found normally at that location.*

Prediction: *As sulfur dioxide is produced by combustion of fossil fuels, the levels of the gas will be highest near main roads, residential and industrial areas. Therefore, only tolerant lichen will be found in these areas. The lowest concentrations of sulfur dioxide will be found on the SW side of the town in the old woodland and farmland areas and so more sensitive lichen will be found here.*

Procedure

Before starting all the students made a copy of the town map on which they marked industrial areas in blue, coloured the main roads in red and all parks, woodland and playing-fields in green. From the town centre they drew eight lines along the main compass bearings (N, NE, E, SE etc.) and made concentric circles from the town centre 0.5 km apart, up to a 5 km radius. Survey points were chosen where the compass lines and circles crossed. Anyone living near an intersection surveyed that location. More distant or difficult points were visited by a group with their teacher. They were told to look on the nearest walls, trees, or hedges to find lichens that were a close match to the photographs of the indicator species.

On return to the laboratory, the results were plotted on the map.

Figure 3.23 The class evaluated their results together. Here are some of their comments

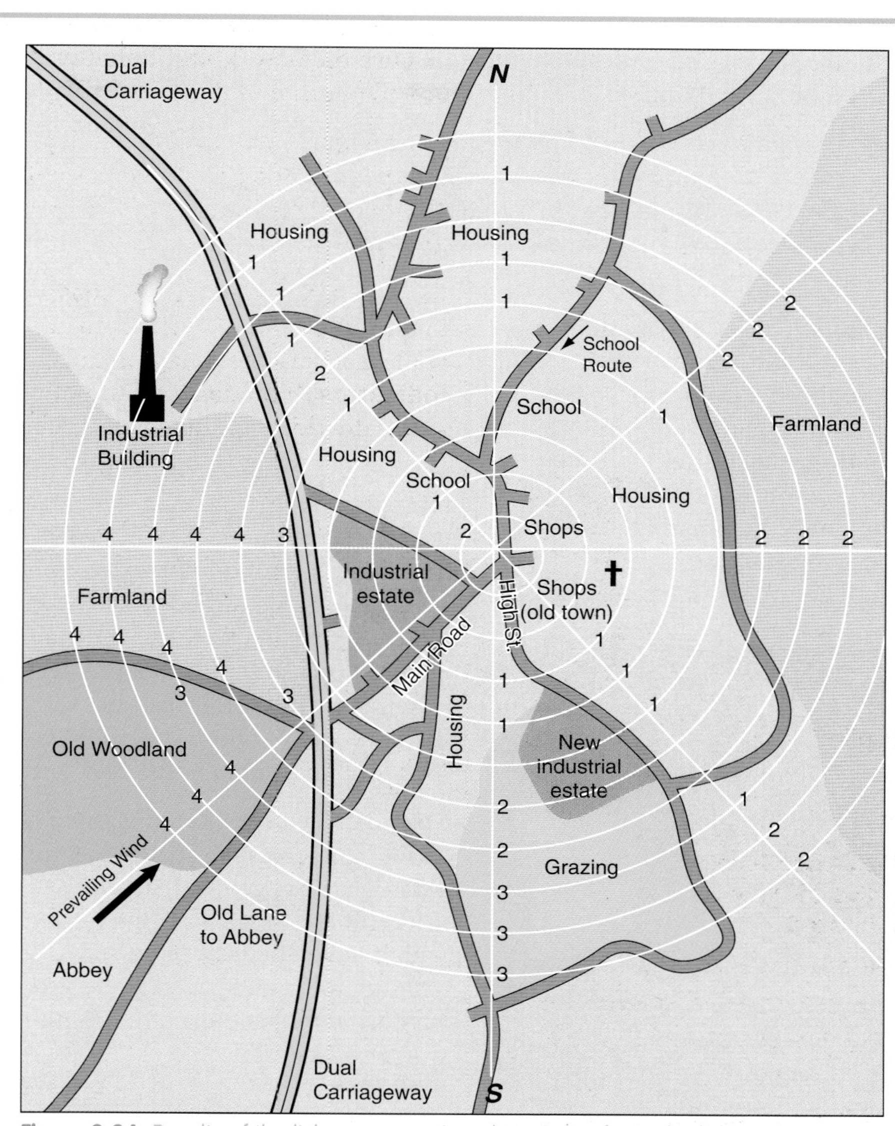

Figure 3.24 Results of the lichen surveys (numbers 1–4 refer to the lichens in Table 3.2)

1 Which gas are lichens most sensitive to and what produces this gas?

2 a) What was the direction of the prevailing wind?
b) How could this affect the distribution of sulfur dioxide?

3 a) Where were the most sensitive lichens found?
b) Did this fit the prediction made?

4 Students reported finding dead and dying lichens in the hedge beside the road running SE from the old town. What does this suggest?

5 a) Where were the least sensitive orange-coloured lichens found?
b) Look at the map and give reasons that support your answer.

6 The churchyard in the town centre had a variety of grey and orange lichens but none was found around a newer church to the east of the town. Suggest why.

7 What instructions were given to students going to a survey point?

8 Using the map, write a conclusion, stating what the survey found.

9. Do the results of this investigation support the prediction made?
10. Using the points made by the group and any others of your own, comment on the **validity** and **reliability** of this investigation.

If you undertake an investigation of this kind, you must follow the risk assessment that your teacher has prepared and check the weather conditions.

Reliable results are results you can trust and that can be reproduced by others.

Valid results are reliable and they answer the question asked. They must be obtained by a fair and unbiased test.

Bioindicators for water oxygenation

Small animals, particularly invertebrates and pollution-sensitive fish, can be used as bioindicators for water cleanliness. Some years ago, there was great excitement when trout were caught in the river Thames after not being seen there for many years. But trout are not usually used as bioindicators in checking water purity; smaller invertebrates are more useful. They can be caught easily in shallow streams (see Figure 3.25). The normal method is kick-sampling, using a D-shaped net to catch the invertebrates disturbed from the stones and plants. The investigator wades across the stream, shuffling the stones with his / her feet. The net is held downstream beside the feet, in the direction of flow, to catch the specimens. For a fair test, the sample is collected by shuffling the same distance and for the same time when each sample is taken. The catch is then emptied into a white tray with a little water in the bottom.

Specimens can be sorted using a teaspoon, pipette and paintbrush. Each different species is recorded as rare, frequent or in large numbers. A clean well-oxygenated stream will have a high biodiversity (many different species). A polluted stream will have a low number of species which have adaptations to survive in low oxygen conditions. An example of such species are chironomid midge larvae, which are red because they have haemoglobin to pick up oxygen in poorly oxygenated water.

Figure 3.25 Students kick-sampling in shallow water

Figure 3.26 A kick-sampling catch being sorted

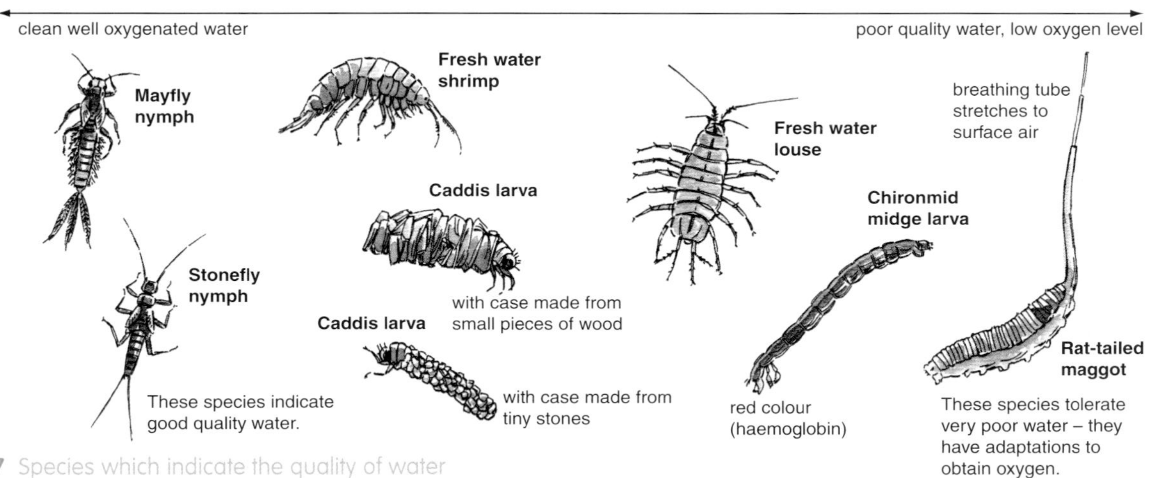

Figure 3.27 Species which indicate the quality of water

Summary

- ✓ Animals must obtain all they need to eat, grow and reproduce from the environment in which they live.
- ✓ Animals have features called **adaptations** which allow them to survive in the environment in which they live. These features enable them to survive at the temperature of the area, find food within the area and locate a breeding site.
- ✓ **Population size** is controlled by **predation** and **competition** for food, water and breeding sites.
- ✓ Plants growing in dry areas have adaptations to reduce water loss, such as thick waxy cuticles, water storage tissue and long roots.
- ✓ Plant populations are controlled by the numbers of herbivores (grazing) and by competition for light, water and minerals.
- ✓ **A balanced ecosystem** changes slowly and the numbers of producers, herbivores and carnivores remain fairly constant.
- ✓ **Biological control organisms** such as ladybirds, insectivorous birds and ground beetles control pest species, such as aphids. Kestrels and owls control mice.
- ✓ The increasing human population has an impact on the environment through the use of more land for buildings, roads, intensive farming and waste disposal.
- ✓ The increasing human population uses more water and raw materials.
- ✓ The increasing human population produces more air pollution, more water pollution, and more waste.
- ✓ **Deforestation** for timber production, arable farming and cattle ranching can have damaging effects on the whole environment including the land, the wildlife and the human population.
- ✓ **Global warming** is the result of increased carbon dioxide and methane levels in the atmosphere.
- ✓ Conclusions linking global warming and greenhouse gas levels require careful measurements over many years, not anecdotal evidence.
- ✓ In environmental situations where many variables are possible, scientists construct computer models to manipulate the data and make predictions for the future. This applies to predictions about global warming.
- ✓ The consequences of global warming are climate changes which result in the flooding of low-lying areas and change in the direction of rain-bearing winds.
- ✓ Action taken by the public to reduce waste, recycle more, and use energy efficiently will benefit society and the environment.
- ✓ **Sustainable development** is necessary to make sure that fuel, food and a pleasant environment are available for future generations.
- ✓ Environmental monitoring can be carried out through chemical sampling, but **bioindicators** of water or air pollution give a longer term view of changes.
- ✓ Sampling of bioindicators requires a well-planned investigation and data collected over many years for comparison.

EXAMQUESTIONS

1. An animal living in the Arctic is likely to have:
   - **A** thick fur, thick fat layer, long thin legs, white fur;
   - **B** thick fur, thick fat layer, short thick legs, white fur;
   - **C** thick fur, thick fat layer, long thin legs, dark fur;
   - **D** thick fur, thin fat layer, short thick legs, white fur. *(1 mark)*

2. The factors which would maintain low population numbers of robins are:
   - **A** cold winters, shortage of nest sites, local cats which hunt, warm spring with many insects;
   - **B** mild winters, shortage of nest sites, local cats which hunt, warm spring with many insects;
   - **C** mild winters, shortage of nest sites, local cats which hunt, cold spring with few insects;
   - **D** cold winters, shortage of nest sites, local cats which hunt, cold spring with few insects. *(1 mark)*

3. Which statement best describes biological indicator species?
   - **A** Living organisms which can survive in a wide range of conditions.
   - **B** Fossil remains that are found in sedimentary rocks.
   - **C** Living organisms that are sensitive to changes in the environment.
   - **D** Living organisms that change colour to suit their background. *(1 mark)*

4. Which definition best describes sustainable development?
   - **A** Development which provides for present needs but may change the environment.
   - **B** Development which provides for future needs by changing the environment.
   - **C** Development which has low environmental impact and provides for present and future needs.
   - **D** Development which has high environmental impact and provides for present and future needs. *(1 mark)*

5. As part of the policy for sustainable development the government proposes 'a reduction in development on "green field sites" and increased use of "brown field sites"'.
   - a) What is meant by the term 'sustainable development'? *(2 marks)*
   - b) How will this proposal benefit the environment? *(3 marks)*

6. Four groups of students investigated the invertebrates in West Brook, a shallow stream. Kick-sampling was used to obtain samples. Groups A and B were sampling on the upstream side of a row of cottages, where the water was clear and the stones on the bottom could be seen easily (Figure 3.28). Groups C and D sampled on the downstream side from the cottages. Here the water was less clear and the stones were covered with a black slimy deposit.

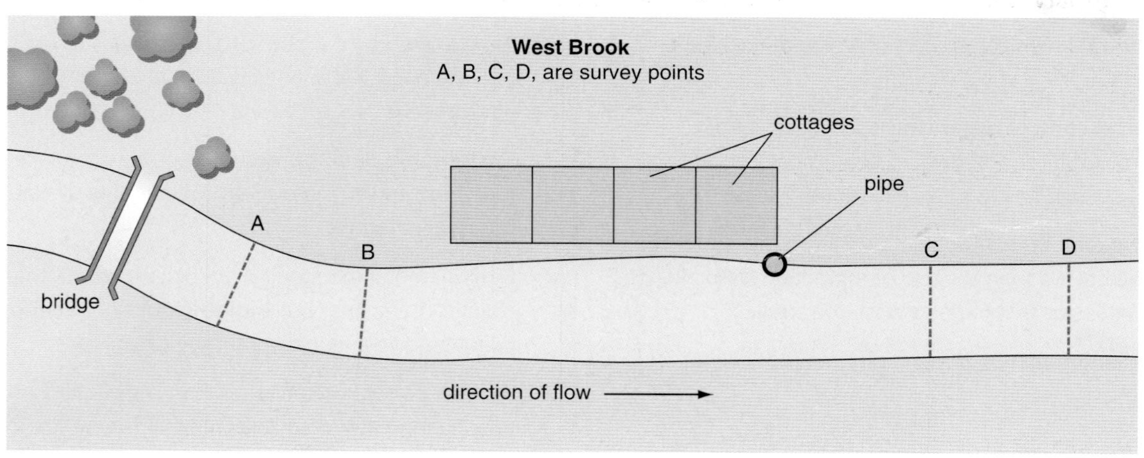

Figure 3.28 A map of the survey sites and cottages at West Brook

Before starting the groups discussed how they would make their investigations a 'fair test'. The results of their samples are shown in Table 3.3.

a) Copy and complete the bottom row in the table and calculate the average number of species in the upstream samples and downstream samples from the cottages. *(3 marks)*

b) What should the four groups do to make 'fair tests' and use their sampling results to compare the two parts of the stream? *(1 mark)*

c) Describe the evidence in the table which indicates that the upstream region was 'healthy'. *(3 marks)*

d) What could have happened by the cottages to change the life in the stream? Give two pieces of evidence which support your suggestion. *(3 marks)*

Organisms	**Oxygenation of water**	**Group A**	**Group B**	**Group C**	**Group D**
Stonefly nymphs	Only survive in well oxygenated water	+++++	+++++		
Mayfly nymphs	Only survive in well oxygenated water	+++	+++++		
Dragonfly nymphs	Survive in oxygenated water		+		
Small fish	Survive in oxygenated water	+			
Freshwater shrimps	Require a reasonable level of oxygenation	+++++	+++++		+
Freshwater lice	Can survive in low levels of oxygenation	+++	+++	+++	+++
Caddis with vegetation case		+	+		
Caddis with stone case			+		
Daphnia		+++++	+++++		
Water snails		+++++	+++++		
Water boatman beetles		+++	+		
Small worms		+++++	+++++		
Midge larvae (red)	Survive at low oxygen levels	+	+	+++++	+++++
Rat tailed maggots	Survive at low oxygen levels			+++++	+++++
Total number of species					

Table 3.3 The results of an invertebrate survey of West Brook
Key: + indicates one specimen, +++ several, +++++ large numbers.

Chapter 4
How can we explain reproduction and evolution?

At the end of this chapter you should:

- ✓ be able to explain how genes are passed from parents to their offspring;
- ✓ know where the genes that carry genetic information are found;
- ✓ be able to describe the differences between sexual and asexual reproduction;
- ✓ understand the different techniques used to produce clones of animals and plants;
- ✓ be able to judge the economic, social and ethical issues concerning cloning and genetic modification;
- ✓ be able to evaluate different theories of evolution;
- ✓ be able to explain why Darwin's theory of evolution is now the most widely accepted;
- ✓ know how fossils provide evidence for the evolution of different organisms;
- ✓ be able to explain how natural selection has led to the evolution of new species;
- ✓ be able to suggest why different species became extinct.

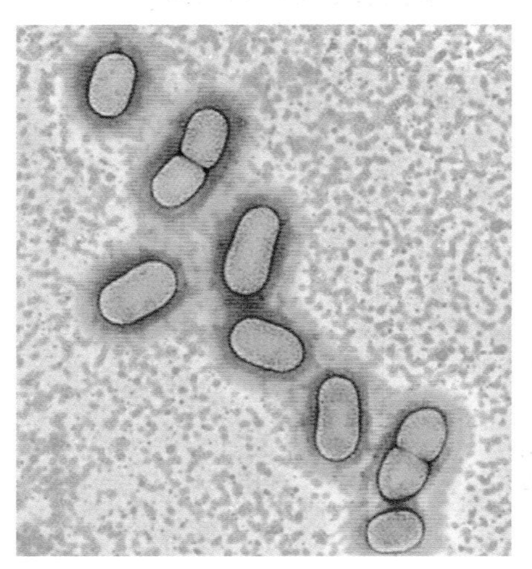

Figure 4.1 In all these photographs characteristics have been inherited from one generation to the next

4.1 How did an Austrian monk help to shape our understanding of inheritance?

Figure 4.2 Gregor Mendel at work

We now know how the features and characteristics that we inherited from our parents are passed on, but this was not always the case. It was only through careful experimentation that scientists gained this understanding. In this section, you will learn about the key developments that led to our current understanding of inheritance.

For many years people believed that the characteristics of parents combined in some way when they had children. For example, if a mother had black hair and a father had blonde hair, the two colours would combine to produce brown-haired children. This was also thought to be true of animals and plants. It was known as the 'blending theory' of inheritance.

An Austrian monk called Gregor Mendel, who was born in 1822, disagreed with this theory. Mendel worked in the gardens of an Austrian monastery and noticed that pea plants had either purple or white flowers. When they produced seeds the new plants that grew only ever had purple or white flowers, never light mauve. He concluded that the parent plants passed on specific features to the offspring plants and not combined features.

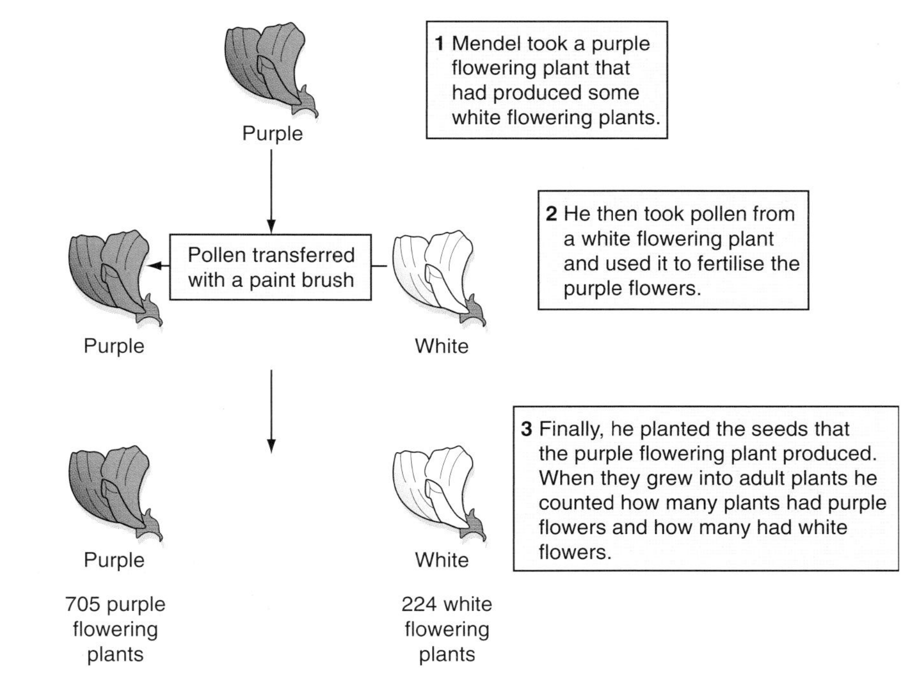

Figure 4.3 A flow diagram showing the procedure Mendel used to investigate inheritance in pea plants and the results he collected

1. What does the word 'inheritance' mean when it is used in science?
2. What prompted Gregor Mendel to carry out his research into inheritance?

3. What did Mendel do to ensure that his findings were reliable?
4. Why was it important for Mendel to obtain reliable results?
5. Explain how Mendel's results showed that the blending theory of inheritance was unsatisfactory.
6. a) Why do you think it took so long for Mendel's ideas to be accepted?
   b) Explain why the work of other people helped Mendel's explanation to replace the blending theory.

Although Mendel had not trained as a scientist, he realised that he needed to carry out controlled experiments. He could then collect reliable evidence to prove his theory. He started by doing some preliminary experiments in which he pollinated purple and white flowered plants together. From this he confirmed that when these two types of pea plants were crossed they produced plants with white flowers and plants with purple flowers but never any plants with colours in between.

For his main experiment, Mendel took a purple flowering plant, that had produced some white flowered plants, and a white flowering plant. He then pollinated these two plants together for several generations and counted the numbers of purple and white flowering plants that were produced.

Mendel analysed his results and discovered that for every three purple flowering plants there was only one white flowering plant, a ratio of 3:1. From his results, Mendel concluded that some inherited characteristics, such as purple flowers, had a stronger influence over the offspring. He called these 'stronger' influences dominant and the 'weaker' influences recessive (see page 142). These findings showed clearly that the blending theory was insufficient to explain inheritance.

Mendel published his findings in 1866 expecting people to appreciate that **inheritance** could be explained much better using his ideas. Sadly, his ideas were largely ignored for 34 years until three other researchers published similar findings based on their own experiments. At the same time, Mendel's work was translated into English and people started to take his ideas seriously. They realised that Mendel provided a far better explanation of inheritance than the blending theory.

Cell structure and inheritance

Mendel pointed out that different characteristics were passed from parents to their offspring. He also showed that some characteristics were dominant and some recessive, but he couldn't explain *how* these characteristics were passed on. It was not until research on cell structure was carried out that the process of inheritance could be explained.

Inheritance is the passing on of characteristics and appearance from parents to their offspring.

A **gene** is a section of a chromosome that carries information for a certain characteristic.

Chromosomes are polymers of DNA. They are found in the nuclei of all cells.

The information that controls inherited characteristics is carried by **genes**. Genes are sections of **chromosomes**, which are long polymer molecules made of deoxyribonucleic acid, or DNA. Chromosomes are found in the nuclei of all cells. There are 46 chromosomes in a normal human cell. These 46 chromosomes carry about 25 000 genes.

Different genes control different characteristics. For example, the flower colour that Mendel studied was controlled by one gene in the pea plant, while the seed colour was controlled by a different gene. Eye colour in humans is also controlled by one gene. These genes are passed on from parents to their children and are carried in the nuclei of sex cells during reproduction. The male sex cells in animals are the sperm cells and in

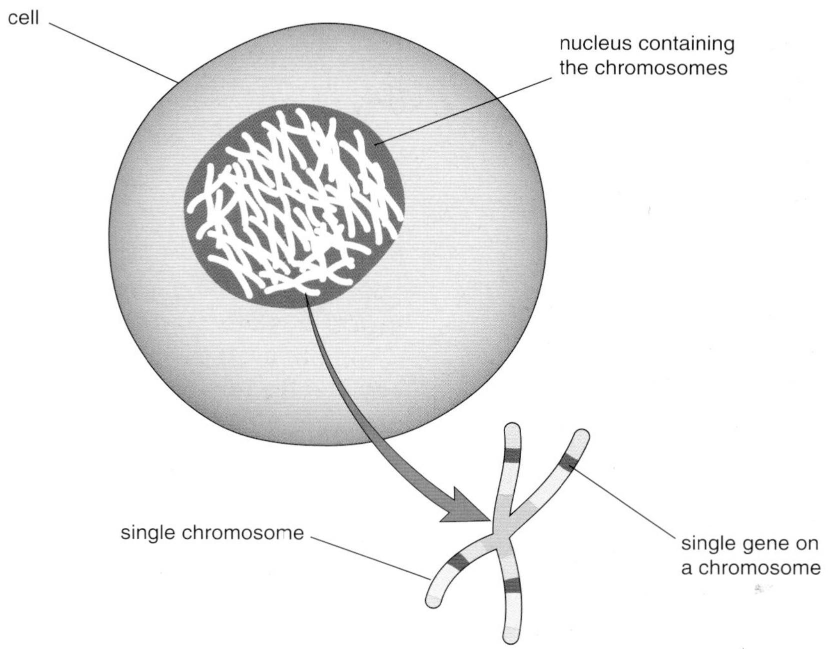

cell

nucleus containing the chromosomes

single chromosome

single gene on a chromosome

Figure 4.4 A diagram showing where genes are found within a cell

Gametes are the sex cells: sperm, egg cells and pollen.

Sexual reproduction is the joining of male and female gametes. It results in variation in the offspring of parents.

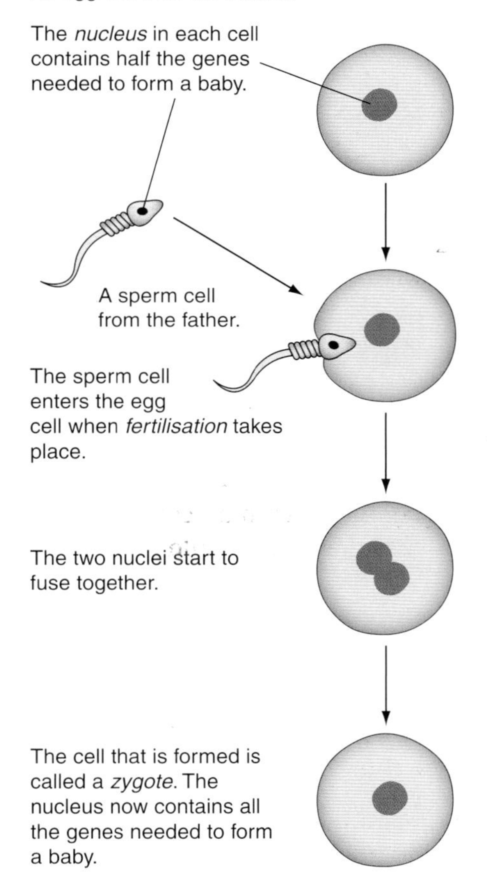

Figure 4.5 A flow diagram showing how genes are inherited by offspring from their parents

plants they are pollen cells. Female sex cells are called egg cells. A general name for sex cells is **gametes**.

Sexual reproduction takes place in animals and plants and involves two parents. During sexual reproduction, a male sex cell fuses (joins) with a female sex cell, to form one new cell. The new cell gets half its genes from one parent and half its genes from the other parent. This new cell develops into a new plant or animal. The offspring's development is controlled by the genes that it has inherited from its parents. Since the offspring has inherited half its genes from each parent, both parents influence its appearance and characteristics.

Have you noticed that children in the same family are often very different in looks and have totally different abilities even though they have the same parents? Scientists call these differences variation. Variation occurs because different genes from each parent can combine when fertilisation takes place. This gives each child in a family a different set of genes unless they are identical twins. It's a bit like selecting five cards from each of two packs. You would almost always end up with two different sets of five cards. As there are about 25 000 genes in a human cell, the combinations are almost endless.

7. What are 'inherited characteristics'?
8. a) Where are genes found?
   b) What do genes do?
9. Why do you think humans need a large number of genes? Explain your answer.
10. How are characteristics inherited by children from their parents?
11. Why does sexual reproduction result in variation of children in the same family?

4.3 How has an understanding of asexual reproduction helped with cloning?

Asexual reproduction produces genetically identical offspring from only one parent. It does not inolve gametes.

Asexual reproduction does not involve gametes and only one parent is needed. This means that all the genes in the offspring come from just one parent. In fact, the genes in the offspring are exact copies of those in its parent. So, **asexual reproduction** produces offspring that are genetically identical to their parents. These genetically identical offspring are known as clones.

Cloning is not as new as you may think. Although there is a lot of media 'hype' and interest in cloning, plant growers have been using cloning techniques for a long time. Some plants and animals reproduce naturally through asexual reproduction, producing clones of themselves. For example, strawberry plants produce runners, which are stems that grow along the surface of the ground. These runners produce roots that grow down into the soil and form a new plant. As the new plant has been produced entirely from one parent, it has exactly the same genes as the parent plant. Bacteria, which are single celled organisms, also reproduce asexually. They copy all their genes and then split in half with a complete set of copied genes going into the new cell. Again, this produces a new individual with identical genes to its parent.

The process of asexual reproduction is used by plant growers to produce cheap, genetically identical new plants by taking **cuttings** from older plants. *Peperomia* are often bought as house plants. New Peperomia can easily be produced from older plants by taking cuttings. This is often carried out in a commercial nursery to mass produce new plants.

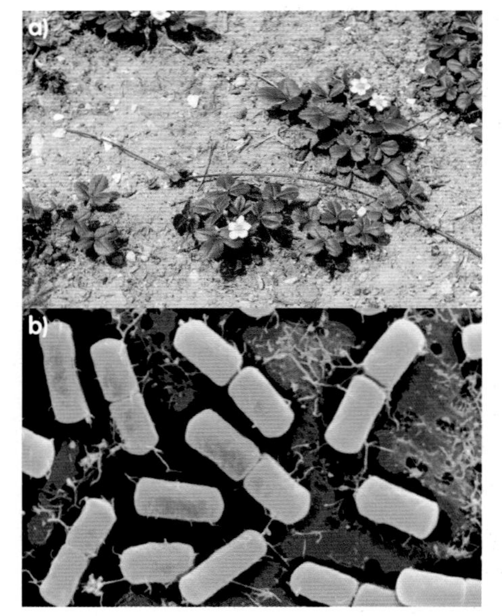

Figure 4.6 Asexual reproduction of a strawberry plant and a bacterium. In each case the offspring is genetically identical to its parent.

Cloning refers to techniques that are used to produce genetically identical individuals.

Cuttings are taken from plants to produce new, genetically-identical plants.

Figure 4.7 A *Peperomia* house plant produced from a cutting

⓬ What are the key differences between sexual and asexual reproduction?

⓭ What is a clone?

⓮ a) What are the benefits to horticultural companies of producing new plants by taking cuttings from older plants?
b) What problems do you think may result from taking cuttings?

⓯ Roots often grow from the cut stem of a plant after a cutting has been taken.
a) What does this suggest about the cells in the stem of the plant?
b) What do you think is the purpose of the hormone rooting powder that cut stems are dipped in?

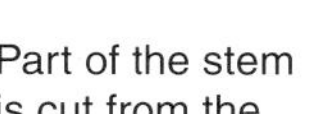

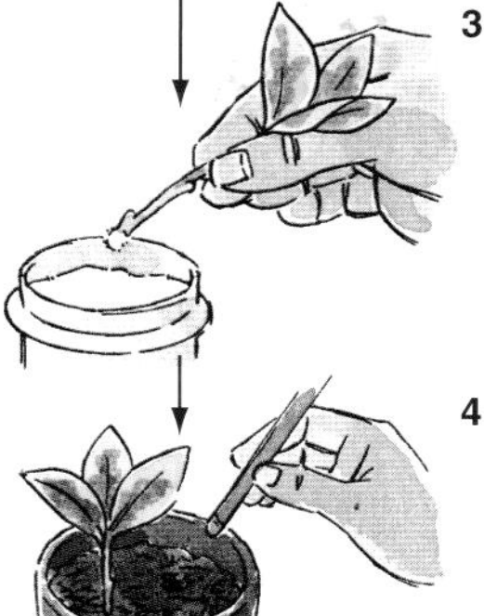

Figure 4.8 A flow diagram showing how a *Peperomia* plant is grown from a cutting

Activity – Modern cloning techniques

Cloning is regularly in the news with headlines such as:

'Cloned cows may be safe to eat.'

'Korean and US scientists claim human cloning breakthrough.'

'UK court ruling means cloning not illegal.'

'After Dolly the sheep comes Snuppy the puppy.'

It is important that you understand the science behind headlines like these in order to form your own views about the media coverage of topics like cloning.

In this activity you will be looking at information about three modern cloning techniques and then considering their potential uses. You will also need to think about the issues that these developments present to society.

Tissue culture or micro-propagation is a way of producing large numbers of plants very quickly. A small number of cells are taken from a 'parent' plant and grown in a medium which is rich in nutrients and plant growth hormones (Figure 4.9).

Embryo transplant is a way of splitting the embryo from a pregnant animal and then transplanting the divided clumps of cells into a number of host mothers (Figure 4.10).

Fusion cell and adult cloning is used to produce exact copies of cells or whole individuals from a single cell (Figure 4.11).

1. Each flow diagram below shows a different cloning technique. For each technique describe how genetic information is passed from the parent plant or animal to the clone.
2. Look carefully at Table 4.1. This shows three benefits and three drawbacks of using tissue culture to clone plants.
3. a) In pairs, discuss the benefits and drawbacks of embryo transplant techniques.
   b) Draw a table showing three benefits and three drawbacks of embryo transplant techniques.
4. a) In pairs, discuss the benefits and drawbacks of using fusion cell and adult cloning techniques.
   b) Draw a table showing three benefits and three drawbacks of fusion cell and adult cloning techniques.
5. The pictures in each of the flow diagrams (Figures 4.9–4.11) show a current application of each cloning technique. Describe briefly another possible application for:
   a) tissue culture;
   b) embryo transplant;
   c) fusion cell and adult cloning.

   You may need to do some research to complete this question. A search on 'cloning' at www.bbc.co.uk will give you lots of examples.

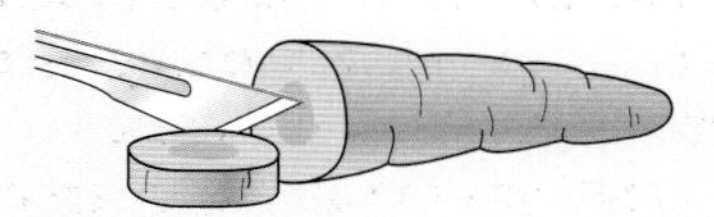

1. A sample of plant tissue is removed from the parent plant.

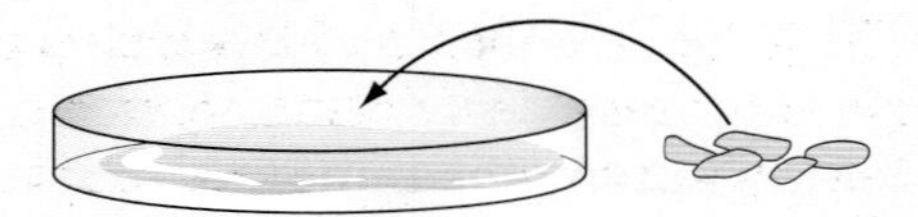

2. The tissue sample is cut into small pieces and placed in a dish of agar jelly containing nutrients and plant growth hormones.

3. Each piece of the tissue sample grows into a clump of cells which develop into small, individual plants called plantlets.

4. The plantlets are transferred into soil or compost where they grow into adult plants.

Figure 4.9 Tissue culture is used to produce large numbers of plants from one parent plant

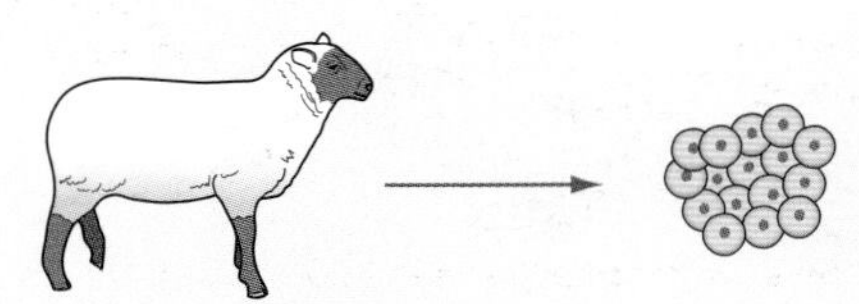

1. An embryo (clump of developing cells) is removed from a pregnant animal.

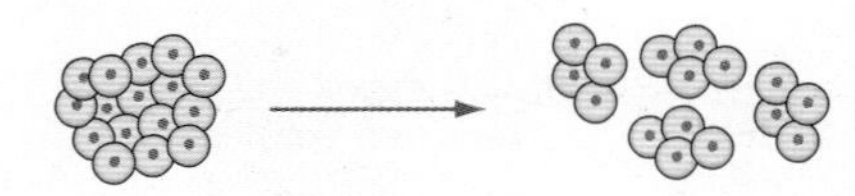

2. The embryo is split into a number of smaller clumps of cells.

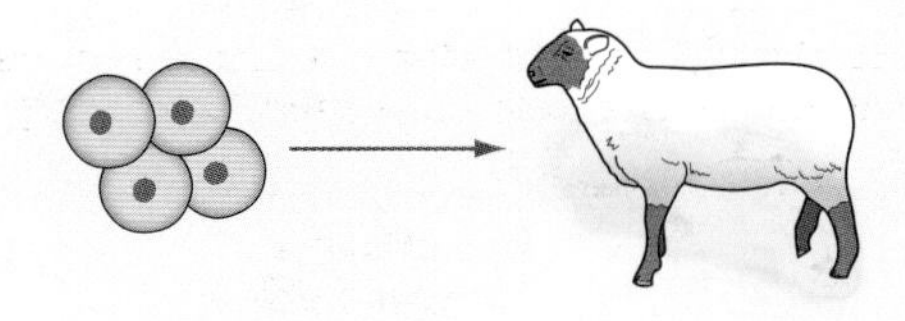

3. Each new embryo is inserted into the uterus of another host mother.

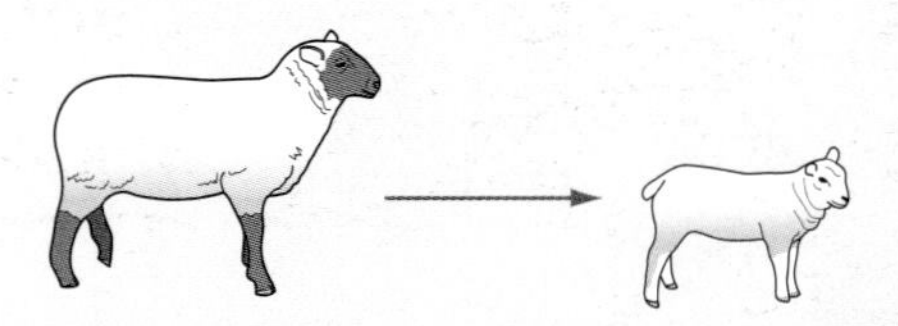

4. Some of the host mothers become pregnant and give birth to their offspring.

Figure 4.10 Embryo transplant is used to implant divided clumps of cells (new embryos) from one embryo into a number of host mothers

Benefits of tissue culture	Drawbacks of tissue culture
A lot of new plants can be grown in a relatively short time	All plants have the same genes, so they will all be vulnerable to the same diseases or pests
Little space is needed, and conditions can be precisely controlled	There is no way that new beneficial characteristics can arise by chance
All new plants inherit the same characteristics	The absence of variation in the plants increases the danger of reducing the gene pool

Table 4.1 Three benefits and three drawbacks of using tissue culture to clone plants

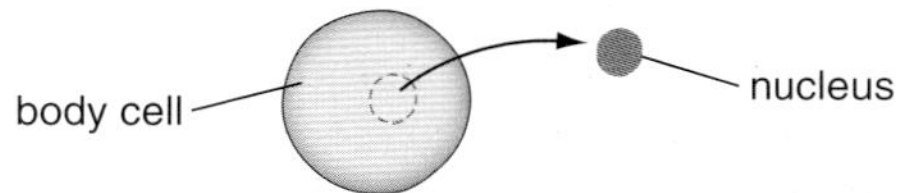

1. A nucleus is removed from a body cell of an adult animal.

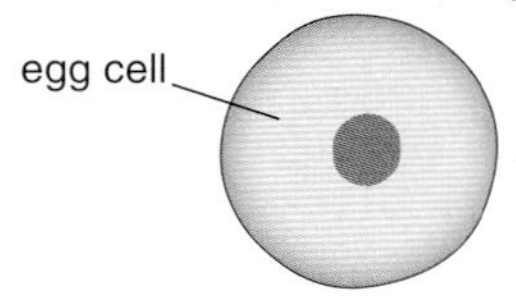

2. An egg cell is removed from another adult female.

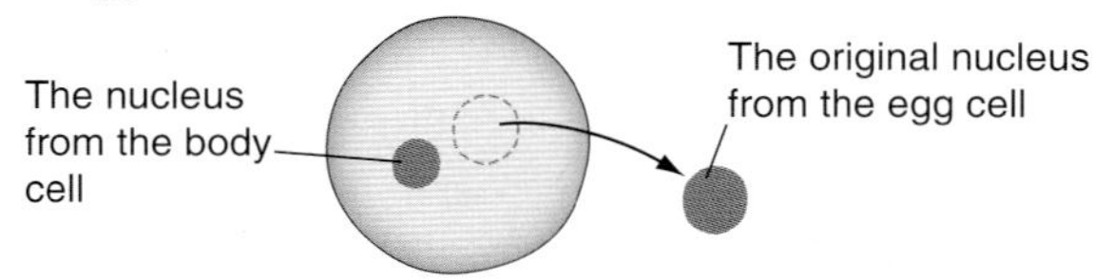

3. The nucleus is removed from the egg cell and replaced with the nucleus from the first adult body cell.

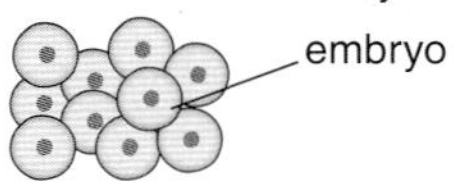

4. The egg cell starts to develop into an embryo.

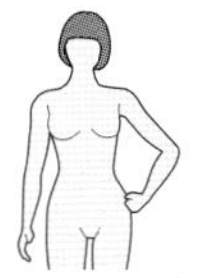
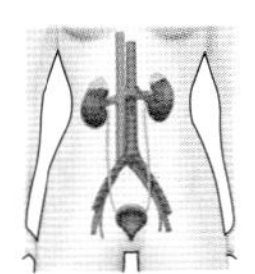

5. The developing embryo is either implanted into another female adult or used to produce a specific body organ.

Figure 4.11 Fusion cell and adult cloning is used to produce cloned animals and human body organs

6 *'Research into cloning humans is wrong and should be banned.'*
This statement is a view held by many people. You are going to prepare a mini-debate about this viewpoint.

- Organise yourself into a group of five. Two people will argue *for* the statement, two will argue *against* the statement and the fifth person will act as a judge.
- Each of the pairs should research and prepare a speech supporting their argument. The judge should research the topic fully and become an expert on cloning.
- Each pair will then deliver their speech to the other pair whilst the judge notes down the main points that are raised.
- The judge will then decide which pair have made the most convincing argument.
- Finally, each judge gives feedback to the whole class stating the main points raised by each pair and giving his / her verdict.

4.4 Understanding genetic engineering

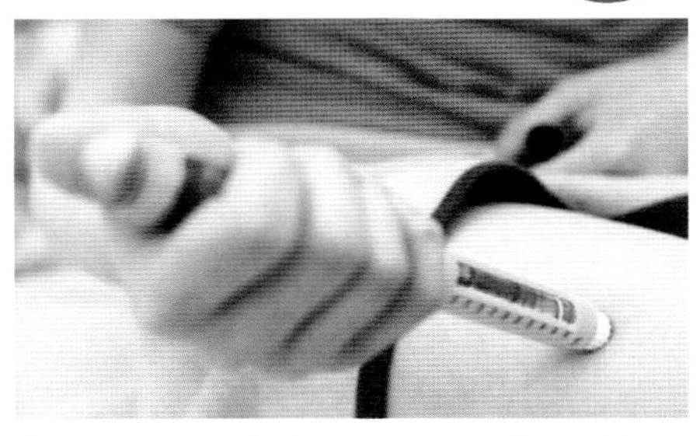

Figure 4.12 A diabetic injecting herself with genetically engineered insulin. In the past, insulin had to be taken from slaughtered animals.

Research involving genetic engineering has been going on for over 40 years. Recent developments show that the techniques could have many useful applications. However, many people are seriously concerned about the possible risks and ethical issues linked to genetic engineering.

Genetically engineering the human growth hormone

Children whose pituitary gland does not produce enough growth hormone suffer from a condition called pituitary dwarfism. They grow very slowly and often reach puberty long after others of the same age. Until the mid-1980s the condition was treated with growth hormone

Figure 4.13 A protest against genetically modified crops

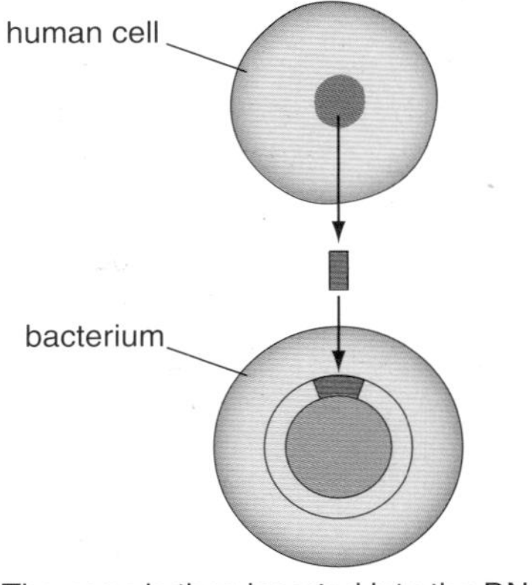

Figure 4.14 A flow diagram showing how genetically engineered human growth hormone is produced

taken from the corpses of unaffected people. This procedure came with a serious risk. Some children treated with the growth hormone developed a disease of the nervous system called CJD. The use of human growth hormone was stopped when this was discovered.

Human growth hormone is now produced by genetic engineering. Chromosomes are extracted from a normal human body cell. The gene that produces the growth hormone is 'cut out' from the chromosomes using an enzyme. The human growth hormone gene is then inserted into the DNA of a bacterial cell. The bacterium then multiplies making lots of bacteria with the human gene. As these bacteria reproduce asexually, each new bacterium has exactly the same genes as the genetically modified bacterium. So, all the new bacteria produce the human growth hormone which can be collected and used to treat the dwarfism. The treatment is expensive but very effective.

Genetically engineering cauliflowers to kill caterpillars

Another application of genetic engineering involves cauliflowers that can kill the caterpillars that try to eat them. Scorpions have a gene in one of their chromosomes that produces the poison they use to kill insects. This gene can be removed from the scorpion's chromosomes with an enzyme. It is then inserted into the nucleus of a fertilised egg cell from a cauliflower. Each time the fertilised egg cell divides it makes a copy of the scorpion's gene, along with copies of its own genes. This results in a cauliflower with the poison producing gene in all of its cells. So, the genetically modified cauliflower now has a poison in its cell sap, which acts as an insecticide. This poison kills any caterpillar that feeds on the cauliflower. This is good news for the farmer, but it leads to a reduction in the local butterfly population.

16. Explain the term 'genetic engineering'.
17. Some parents of children suffering from pituitary dwarfism refused the treatment with growth hormone from human corpses. Suggest two reasons why they refused this treatment.
18. Suggest two benefits of using genetically engineered human growth hormone instead of growth hormone from human corpses.
19. The poison from genetically modified cauliflowers does not harm humans when it is eaten. Why do you think people still refuse to eat the genetically modified cauliflower?
20. Table 4.2 lists a number of applications of genetic engineering. Copy and complete the table adding columns stating whether you agree or disagree with each application and giving one reason for each of your decisions.

Application of genetic engineering
Producing human insulin from genetically modified bacteria
Producing disease resistant rice to increase its yield in developing countries
Producing chickens with four legs and no wings to increase meat production
Parents being able to choose the sex of their baby
Developing plants that can produce plastics to reduce the use of crude oil

Table 4.2

How do new species of plants and animals evolve?

Figure 4.15 Some of the species of animals and plants found on the Earth today

We live on a planet with a fantastic variety of different animals and plants. Scientists have estimated that there are between 2 million and 100 million different species of plants and animals on the Earth today, but they don't know the exact number. How did all these different animals and plants come to be living on Earth? Have there always been similar plants and animals on Earth? Were the present plants and animals like their predecessors (forerunners) or very different from them? Although there are answers to some of these interesting questions, scientists cannot be certain how life began on Earth.

Most scientists believe that the plants and animals alive today have developed from forerunners that lived in the past. These forerunners were not the same as today's species. Changes have taken place over many generations. Scientists say that the present plants and animals have **evolved** from their forerunners and they call the overall process **evolution.** This evolution has resulted in species that are adapted to survive in the environments in which they live. No one has actually seen these changes taking place because the process takes so long. But there have been plenty of theories that try to explain evolution.

Activity – Theories for evolution

In this activity you will be considering five different theories for the evolution of plants and animals. Think about the way in which the theories conflict with each other and try to decide which one offers the most convincing explanation.

Creationism is based on the idea that every species was created separately by God and that species do not change through time. This was the view held by the Christian Church in the seventeenth century and strongly reinforced by Archbishop James Ussher. Ussher studied the Bible and calculated that all life had been created in 4004BC. According to his theory, all species alive today were created about 6000 years ago in the same form as they are now. As plants and animals reproduced, there were no changes from generation to generation.

A second theory of evolution was proposed by the French zoologist, George Leclerc **Buffon**. In the late eighteenth century, a number of European scientists began to question the creationism idea that species

Figure 4.16 Archbishop James Ussher (1581–1656) believed strongly in creationism

had never changed. Buffon actually suggested that the Earth was at least 75 000 years old and that certain species had changed in that time. He explained that these changes were sometimes influenced by the environment and sometimes happened by chance. He even suggested that humans and apes were related. Buffon's anti-creation ideas were kept quiet, and simply recorded in a book called *Histoire Naturelle* which sold in only limited numbers.

Figure 4.17 George Leclerc Buffon (1707–1788) was the first scientist to suggest that species might evolve over thousands of years

In the early nineteenth century, a biologist called Jean Baptiste **Lamarck** published his theory of evolution. He suggested that microscopic organisms could form from non-living material and then gradually evolve into more complex species, eventually resulting in humans. Lamarck believed that these changes happened during an animal's lifetime. The changes that an animal developed were then passed on to its offspring. For example, he explained that herons had evolved long legs because individual herons had stretched their legs to stay dry when wading in water. These herons would then pass on this characteristic to their offspring who would, in turn, stretch their legs even more and produce offspring with even longer legs. He also used this theory to explain why giraffes have such long necks.

Lamarck's theory was severely criticised by another French scientist called George **Cuvier.** Cuvier argued that changes to an individual animal during its life could not be passed on to its offspring. For example, a person who trains hard will develop bigger muscles but will not then have children that grow up having bigger muscles. Cuvier did however agree with Lamarck that there had been different species in the past. Cuvier's theory was called **catastrophism.** It suggested that sudden environmental changes, such as earthquakes, killed whole species. After catastrophes such as this, new species slowly took their places. Cuvier based his ideas on fossil records.

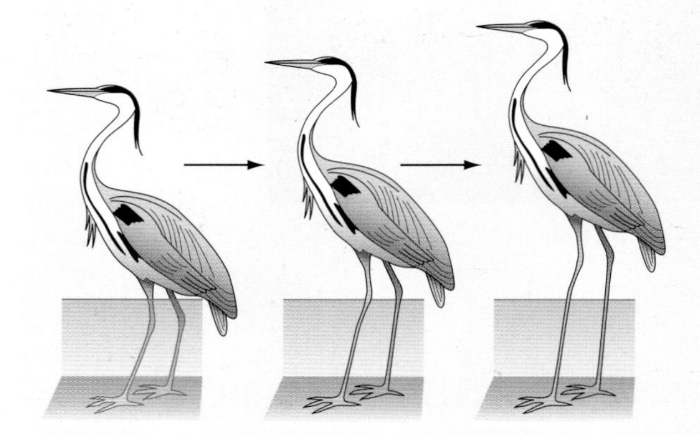

Figure 4.18 Lamarck's explanation for the heron's long legs

The final theory for you to consider is that proposed by the Englishman, Charles **Darwin**, in the nineteenth century. Darwin studied theology but was more interested in biology. He joined the crew of a ship called the Beagle on a five-year voyage to Africa, South America and Australia. During this trip, Darwin studied the wildlife on the Galapagos Islands, which are 1000 kilometres from the west coast of South America. He found that the islands had species of animals that were not found anywhere else in the world. For example, he found 13 species of finches on the islands whereas only one of these species lived on the mainland of South America. The different species of finch on the Galapagos Islands had differently shaped beaks that suited the food they ate. For example, one species had a very strong beak for cracking nuts while another, living on a different island, had a long thin beak for picking up insects.

Darwin suggested that the different finches on the islands had evolved from a single species that had migrated from the mainland. He explained that within a population of finches there were always some variations in features, such as beak size and shape. Individual birds with features that better suited their environment were more likely to survive and pass on these features to their offspring. Darwin found a similar pattern to the finches in other species on the Galapagos islands.

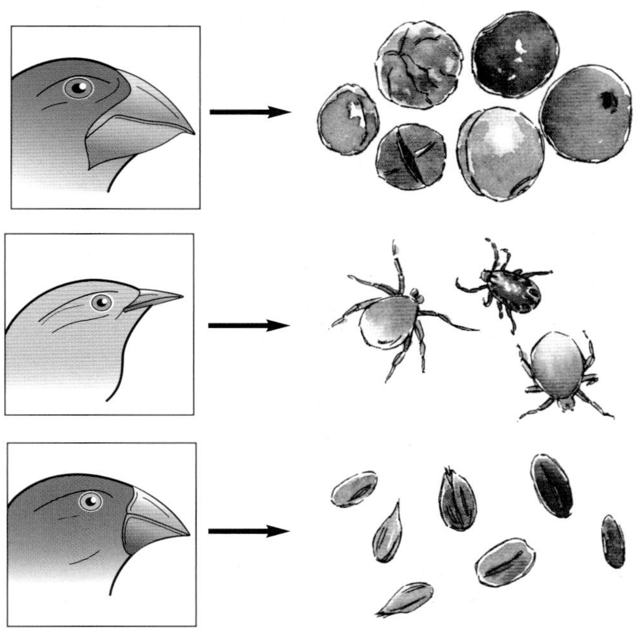

Figure 4.19 Three finches found on the Galapagos Islands and the food they eat

1. Copy and complete Table 4.3 summarising the main points in each theory. The first one has been done for you.

Theory or scientist	Main points
Creationism	• God created all species alive today about 6000 years ago • Species have not changed over time
Buffon	
Lamarck	
Cuvier / catastrophism	
Darwin	

Table 4.3

2. Why do you think Buffon chose not to publish his ideas?
3. Explain the problems with Lamarck's theory, using your understanding of genes and inheritance. If you need help, read through Section 4.2.
4. Darwin's theory of evolution is now widely accepted.
   a) What are the main differences between Darwin's ideas and the other theories?
   b) Why do you think Darwin's ideas give a better explanation of evolution?

Figure 4.20 This drawing of Darwin's head on a monkey's body was printed after he published his theory of evolution

5. Darwin's ideas were only gradually accepted. Look at Figure 4.20. Why do you think a cartoon like that in Figure 4.20 was printed?
6. Creationism reflected a religious belief about life on Earth. Today, Christians still believe that God created life and many of them also accept Darwin's theory of evolution. Discuss in pairs how Christians can believe that God created life on Earth and also accept Darwin's theory of evolution.

4.6 Evidence for evolution and natural selection

Darwin formed his theory of evolution from his observations on the Galapagos Islands and elsewhere. Today we have much more evidence to support his theory. The theory of evolution states that all species have evolved from simple life forms which first appeared on the Earth 3000 million (3 billion) years ago. So how do we know what the forerunners of today's species were like? How did we get our evidence for evolution?

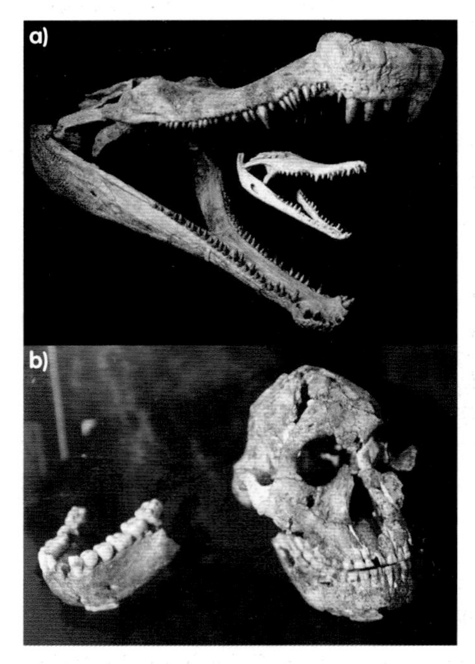

Figure 4.21 Fossils of a) a crocodile's jaw and teeth from 200 million years ago and b) the bones of *Orrorin tugenensis* from 2 million years ago

Most of our evidence for evolution comes from fossils. Scientists study fossils to find out about the animals and plants of the past. Fossils show how much or how little organisms have changed. Fossils of crocodile jaws and teeth show that they have evolved very little over the last 200 million years (Figure 4.21). In comparison, the earliest fossils of our human ancestors, such as *Orrorin tugenensis*, which was the size of a chimpanzee, show that humans have evolved a great deal in just 2 million years.

Fossil records show how the modern whale could have evolved from a land-based mammal (Figure 4.22). The original land-based mammal from which whales have evolved was called Pakicetidae, about the size of a wolf. Millions of years ago Pakicetidae started to go into water to find a more abundant source of food. Within any population of the animals, some would be better swimmers. These animals may have had wider feet or have been more streamlined. As a result, these individuals could catch more food and were more likely to survive and reproduce. The genes that controlled these features would be passed on to their offspring who would inherit wider feet and more streamlined bodies. These individuals would be the best swimmers in the next generation. Over millions of generations of gradual change, it is possible that Pakicetidae was transformed into the modern whale. Of course, this is only a theory but it is supported by fossil records.

21. What information about life on Earth can be obtained by studying fossils?
22. Why is it important to have a series of related fossils when studying evolution?
23. Fossil records for the evolution of whales provide evidence to support the theory of evolution, but they do not prove it. Explain why this is so.

24 It has been suggested that today's horses evolved from an ancestor the size of a fox. Write the following sentences in the correct order to explain how the horse may have evolved by natural selection.

- Gradually, over many generations, taller horses with the toes joined together were produced.
- Some *Eohippus* had slightly longer legs than others and could run faster.
- Some *Eohippus* also had toes that were closer together, which helped them to run even faster.
- Eventually the modern horse evolved with long legs and hooves instead of toes.
- The individuals that survived were more likely to reproduce and pass on their characteristics to their offspring.
- Sixty million years ago, a small mammal, *Eohippus,* existed that had three toes and short legs.
- These individuals were better at avoiding predators and were more likely to survive.

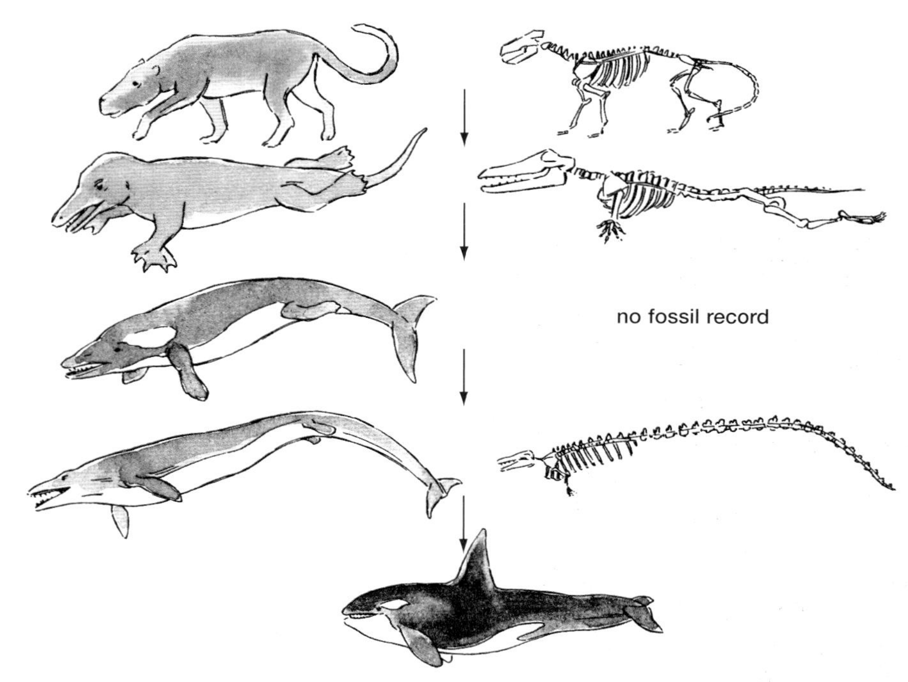

Figure 4.22 Pictures showing how the whale may have evolved from a land-living mammal

The sequence of events that leads to evolution of a species is called natural selection. Natural selection can be split into five key steps.

1. The individuals within a species vary because of differences in their genes. These variations result from the inheritance of different genes or from the occasional mutation (change) of genes.
2. Individuals with the characteristics that best suit their environment are more likely to survive.
3. These individuals are also more likely to reproduce successfully.
4. The genes which have enabled these individuals to survive are then passed on to the next generation.
5. This process is repeated over many generations leading to a new species.

Why do species become extinct?

An animal becomes extinct when there are no more living members of its species. Extinction is often part of evolution. As a new species evolves, the old one dies out. Sometimes a species becomes extinct without a new species evolving. The main causes of extinction are:

- changes to the environment;
- new predators;
- new diseases;
- new competitors.

Figure 4.23 The dodo became extinct around 1700.

It is difficult to know the cause of an extinction when it occurred so long ago. The dodo (Figure 4.23) was a flightless bird that lived on the

25 Why are we still not sure what caused the dodo to become extinct?

26 In the last one hundred years, humans have been responsible for a large increase in the number of species that have become extinct.
  a) Which human activities have caused this increase in extinction rates?
  b) What problems could this loss of species cause in the future?

island of Mauritius. It finally became extinct around 1700 after humans had settled on the island. For many years, scientists thought that the dodo became extinct because the people that moved to the island hunted and ate the dodo. In other words, humans were a new predator to the dodo. More recently, however, records suggest that the dodo was not widely eaten by the settlers.

When humans moved to Mauritius, they cut down most of the forests to make space for their homes. In doing so, they destroyed the habitat in which the dodos lived. As the dodos were flightless, they could not easily migrate to a different island. So, changes in the environment may have resulted in extinction of the dodo.

The settlers brought a number of animals with them that were new to Mauritius. These included monkeys, pigs and rats. These animals ate the same food as the dodos and destroyed their nesting sites. The new animals competed with the dodos for habitat and food, which might also explain why the dodos became extinct.

Summary

- ✓ Some characteristics are passed on from parents to their offspring. This is called **inheritance**.
- ✓ The information that controls inherited characteristics is carried by **genes** which are passed on in the sex cells (**gametes**).
- ✓ A gene is a section of a **chromosome**. Chromosomes are found in the nuclei of all cells.
- ✓ **Sexual reproduction** is the joining of a male and a female gamete. Sexual reproduction involves the mixing of genetic information from two parents and leads to variety in their offspring.
- ✓ **Asexual reproduction** only requires one parent and does not involve the joining of gametes. Offspring produced by asexual reproduction from one parent are genetically identical and show no variation.
- ✓ Genetically identical individuals are called **clones**.
- ✓ New plants can be produced quickly and cheaply by taking cuttings. This is a type of cloning.
- ✓ Modern cloning techniques include tissue culture, embryo transplants and fusion cell and adult cloning.
- ✓ It is important to be able to make informed judgements about the economic, social and ethical issues concerning the use of cloning and genetic engineering.
- ✓ Genetic engineering can be used to transfer genes from the cells of one organism to the cells of other organisms.
- ✓ Genes can be transferred to the cells of plants or animals to introduce new characteristics to the organism.
- ✓ There have been a number of conflicting theories to explain evolution.
- ✓ Darwin's theory of evolution is now the most widely accepted. Darwin's theory explains how the evolution of new species occurs by natural selection.
- ✓ The theory of evolution states that all species have evolved from simple life forms which first appeared more than 3 billion years ago.
- ✓ Fossils provide us with strong evidence for the theory of evolution.
- ✓ The extinction of a species may be caused by: changes to the environment, new predators, new diseases or new competitors.
- ✓ It is not always easy to pinpoint what caused a particular species to become extinct.

EXAMQUESTIONS

1. Young rabbits, like these, often look like their parents. This is because information about their appearance, for example fur tone, is handed down from parents to their offspring.

Copy and complete the paragraph below using the appropriate words from the box.

body	chromosomes	clones	
cytoplasm	genes	nucleus	sex

Information is passed from parents to their offspring in ____________ cells. Appearance is controlled by __________________. The structures which contain information about a large number of characteristics are known as the ____________________. In the cell these structures are found in the ____________________. *(4 marks)*

2. Humans reproduce by sexual reproduction. Michael and James are brothers, but they have very different features.

a) State two ways in which sexual reproduction is different from asexual reproduction. *(2 marks)*
b) Explain why Michael and James have different features. *(3 marks)*

3. Read the piece of text below carefully.

A woman from Texas ordered a genetic replica of her cat, Nicky, when he died aged 17. The new cloned cat, named Little Nicky, who cost his new owner $50 000, was created using DNA taken from his namesake, Nicky. 'He is identical. His personality is the same' said the owner. During an interview the new owner asked that her surname and hometown were not made public. She feared that she may become a target for groups that oppose the use of cloning techniques.

a) Explain why Little Nicky is exactly the same as Nicky. *(3 marks)*
b) Little Nicky's owner is concerned that she may become a target for anti-cloning groups.
   i) Give one example of a group or organisation that may disagree with the cloning of pet animals. *(1 mark)*
   ii) Why do you think such groups disagree with these cloning techniques? *(1 mark)*

4. a) Provide an explanation for the theory of evolution. *(2 marks)*
b) The text below provides a possible explanation for the evolution of the long legs of wading birds, by Lamarck.

When an animal undergoes change during their lifetime and then mates, they pass this change on to their offspring. For example, wading birds did not always have long legs. They gradually developed them after their ancestors started to feed on fish. As they walked into deeper water, they would stretch their legs to prevent their bodies from becoming wet, causing their legs to lengthen. Their new trait of longer legs would be passed on to their offspring, who would also stretch their legs. Over time, the legs of these wading birds became longer and longer.

Darwin would have provided a different explanation for the evolution of the long legs of these birds. What are the main differences between his explanation and that of Lamarck? *(3 marks)*

Chapter 5
How do substances get into and out of cells?

At the end of this chapter you should:

- ✓ know the names and be able to explain the function of the different parts of animal and plant cells;
- ✓ be able to describe the differences between animal and plant cells;
- ✓ be able to identify a range of specialised cells;
- ✓ be able to relate the structure of a specialised cell to its function;
- ✓ be able to explain how substances move in and out of cells by diffusion;
- ✓ know how water moves in and out of cells by osmosis;
- ✓ understand how osmosis is affected by the concentrations of the solutions inside and outside a cell.

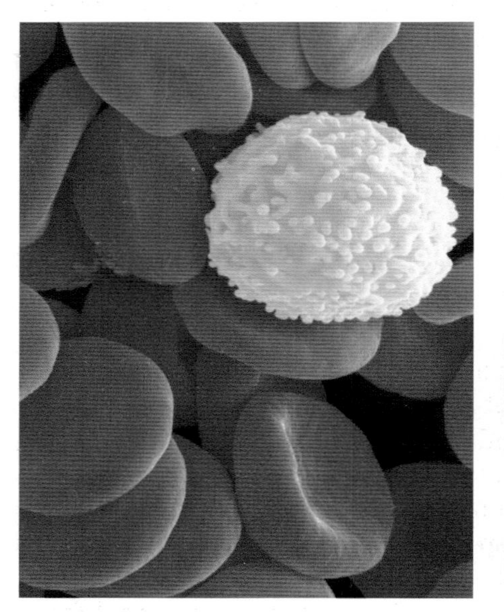

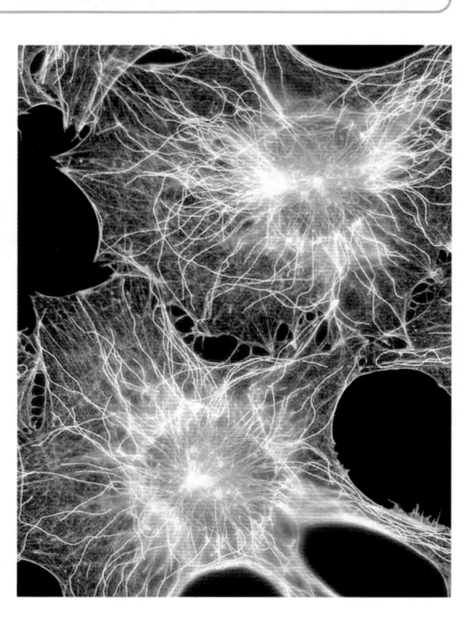

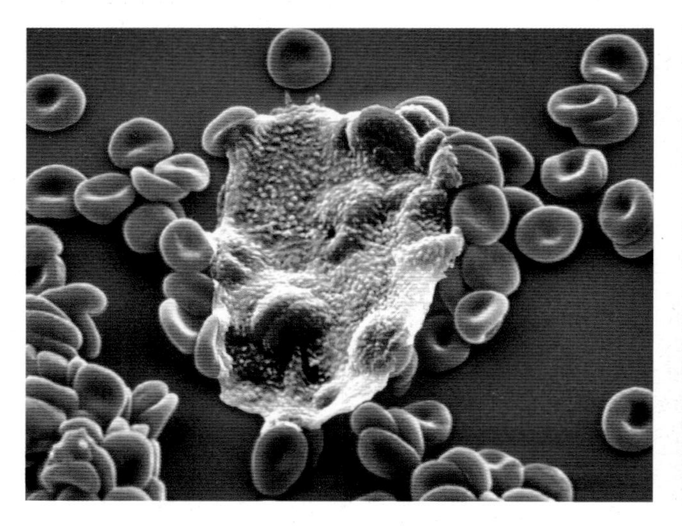

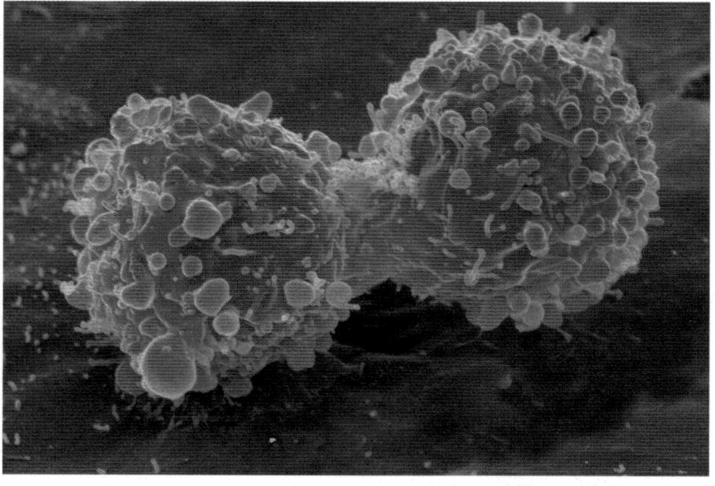

Figure 5.1 Cells are essential for life. They control all the processes in your body that keep you alive and are responsible for reproduction. The cells in the pictures are all very different and this allows them to do different jobs.
Top: Red and white blood cells; pollen cells, fibroblast making new connective tissue. Bottom: macrophage killing red blood cells, cancer cells.

How do cells perform different functions?

We now understand that all living things are made up of cells, but this was not always the case. In 1665 a scientist called Robert Hooke used a primitive microscope to discover that samples from a cork tree were made up from repeating structures. He called these **cells**.

Hooke's discovery was only the start of what is now known as cell theory. Although he observed cells in one type of plant, he did not link cells to other species and had no idea what was inside cells. In 1831 improvements in microscope technology allowed a scientist called Robert Brown to observe nuclei inside a number of different cells. Later in the same decade two German biologists, Theodor Schwann and Matthias Jakob Schleiden, carried out more research and linked cells to all plants and animals. Schwann published these findings without any acknowledgment of Schleiden's or anyone else's contributions. He was correct to suggest that cells are the unit that make up all living things. However, he also suggested that new cells formed out of nothing. This was proved to be wrong by Rudolph Virchow in 1858. Virchow proposed the theory, still in place today, that all cells are formed from other cells.

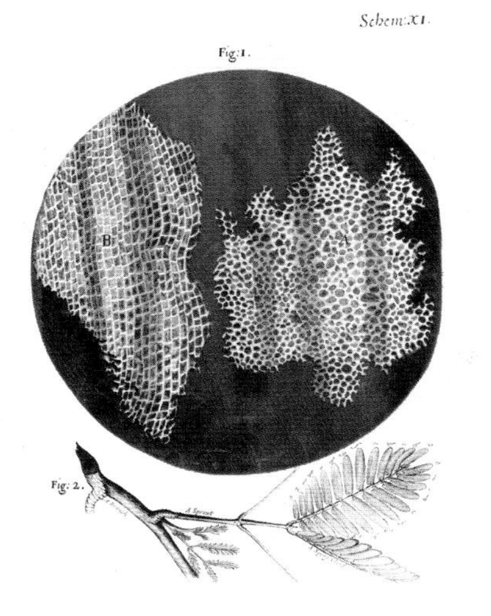

Figure 5.2 Robert Hooke and the drawing he made of the cork cells he could see with his microscope

What are the functions of different parts of a cell?

Modern microscopes now let us see an incredible amount of detail about the structure of cells, details that Robert Hooke's microscope simply could not display. We can now identify different parts of cells that each carry out different functions. From our observations we have discovered that all cells have some parts in common, such as a cell membrane. However, parts of some cells have specialised shapes so they can carry out particular functions. For example, you can see a fibroblast cell in Figure 5.1, making new connective tissue. This process is essential for the body to heal after an injury. There are also some fundamental differences between the cells that make up animals and those found in plants.

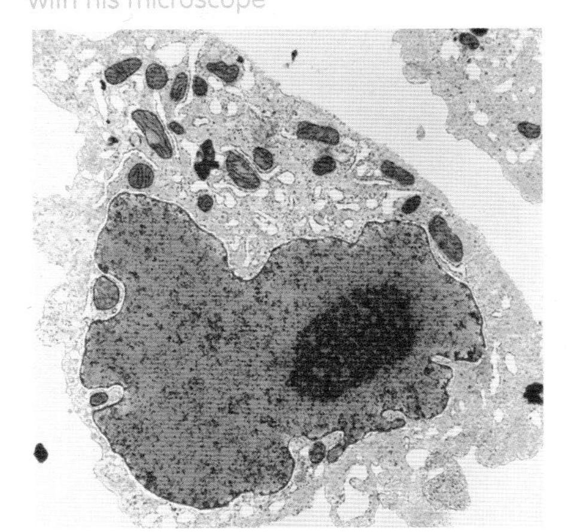

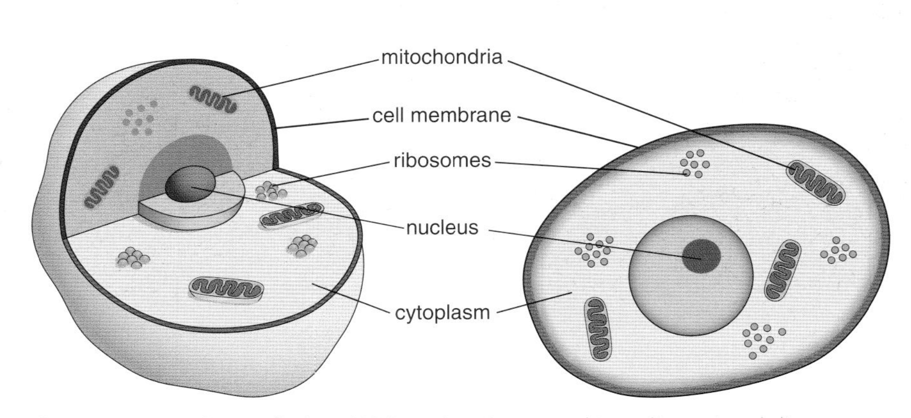

Figure 5.3 An animal cell shown as it is seen with a microscope (magnified × 4800) and as three- and two-dimensional diagrams

1. Cells are often described as the 'building blocks of life'. What does this mean?
2. Robert Hooke observed cells but did not identify anything inside them. What was limiting his research?
3. Why do you think that Robert Brown's discovery was important in developing a better understanding of cells?
4. No one scientist can claim that they came up with cell theory. However, Schwann did not acknowledge anybody's contributions when he published his findings. What reasons do you think he had for doing this?

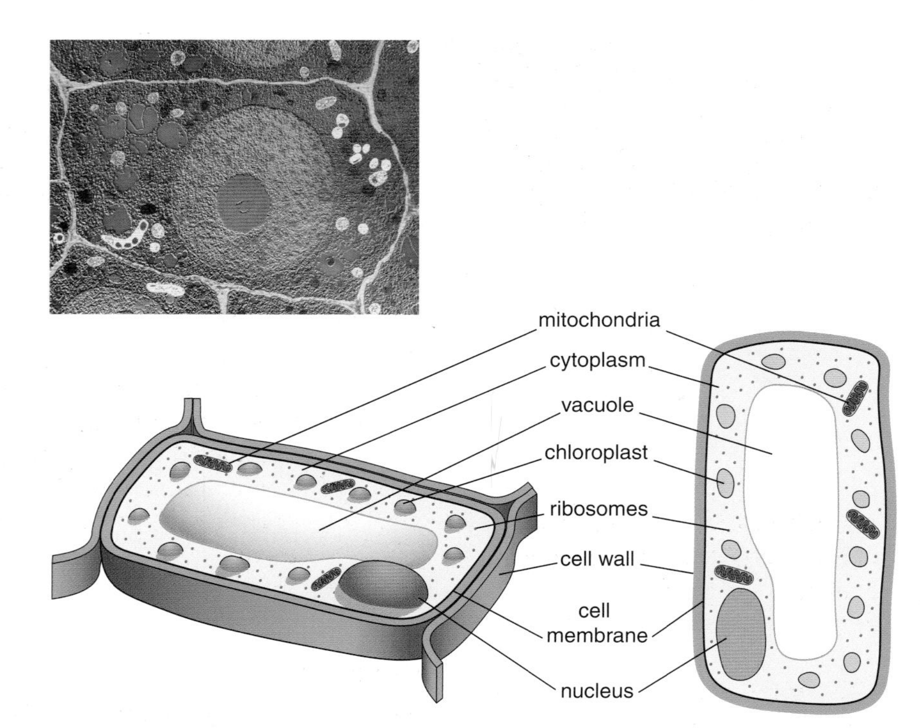

Figure 5.4 A plant cell shown as it is seen with a microscope (magnified × 2800) and as three- and two-dimensional diagrams. Note the cell wall, vacuole and chloroplasts that are not found in animal cells.

Proteins are long chains of amino acids that form the basis of many body structures such as skin, muscle and hair. They also form chemicals in the body such as enzymes.

Similar cells group together to form body tissue. For example, your skin is made up of layers of skin cells and your muscle is formed from muscle cells packed together. New tissue forms when cells divide. The cells make exact copies of themselves. This includes making copies of all the cell parts you can see in Figures 5.3 and 5.4. Each of these parts carries out a job. This enables the cell to contribute to the function of the tissue it is part of. Now you need to look at the function carried out by each part of a cell.

Nucleus

The nucleus of a cell contains an individual's genes. These carry out two functions:

1. They can make copies of themselves when the cell divides to make a new cell.
2. They synthesise (make) **proteins**, such as enzymes.

Enzymes control all the reactions that take place in the cell. These reactions include respiration, photosynthesis and protein synthesis. Enzymes even control cell division when it takes place. Some of these enzymes pass out of the cell membrane to control reactions outside the cell. For example, amylase is an enzyme produced by cells in your salivary glands. Amylase passes out of the cells and digests starch into sugar in your mouth and oesophagus. Ideas about enzymes are developed in Chapter 7.

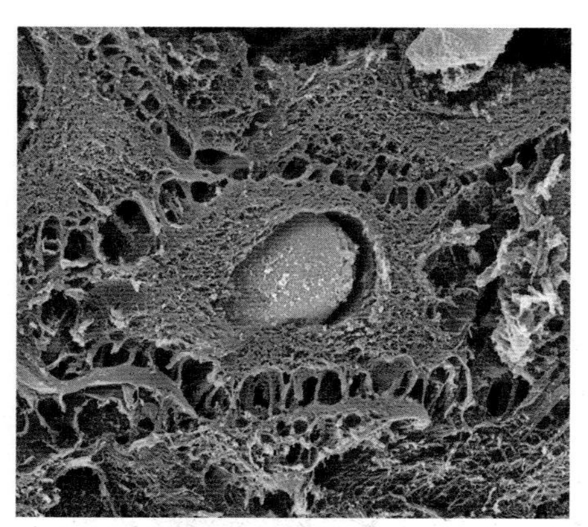

Figure 5.5 The nucleus of a human skin cell

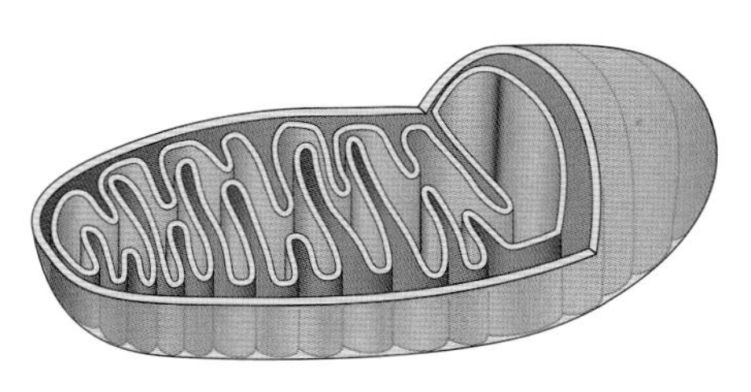

Figure 5.6 A human mitochondrion

Cytoplasm

The cytoplasm is mostly water and fills the cell membrane. It is where most of the cell's chemical reactions take place. When a cell is making an enzyme, this happens in the cytoplasm. The cytoplasm also stores the dissolved raw materials, such as amino acids and glucose, needed for all these reactions.

Mitochondria

Your whole body needs energy to keep warm, grow and move around. This energy gets into your body in your food where it is stored in the chemical bonds in the food. **Mitochondria** release the energy from glucose molecules during a process called respiration. Respiration is a series of chemical reactions that take place on the surface of the channels you can see inside the mitochondrion shown in Figure 5.6. These channels have a large surface area so that respiration can take place at a fast rate.

Mitochondria are the parts of a cell responsible for releasing energy from glucose by respiration.

Ribosomes put amino acids into the correct order to make a specific protein.

Ribosomes

You have already seen that the genes in the nucleus make proteins in the cytoplasm, such as enzymes which control all the chemical reactions in your body. **Ribosomes** are structures that control the making of proteins. They join together the amino acids into the long chains that become proteins.

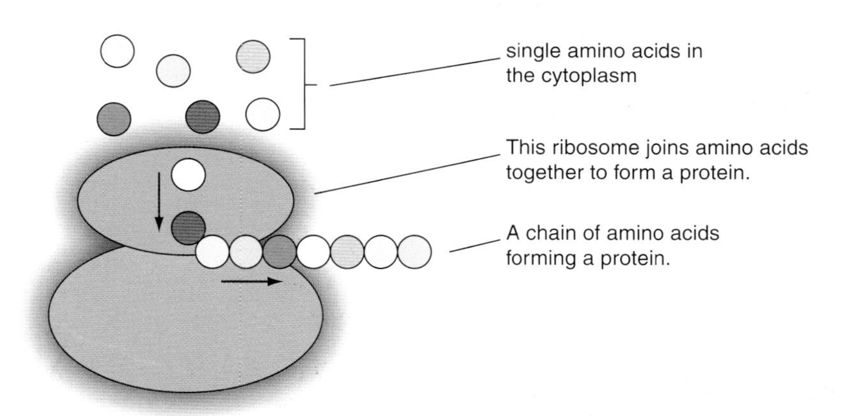

Figure 5.7 A ribosome forming a protein chain from amino acids in the cytoplasm

Cell membrane

The cell membrane surrounds the cytoplasm. Anything that has to get in or out of the cell has to pass through the cell membrane. For example, the glucose used in the mitochondria for respiration enters the cell through the membrane, and insulin produced in pancreas cells leaves these cells through the membrane. The cell membrane actually controls the movement of many of these substances in and out of the cell.

Cell wall – found only in plant cells

Plant cells have a rigid cell wall around the cell membrane that strengthens the cell and keeps its shape. It is made from a chemical called cellulose. Without cell walls plants would not be able to stay upright. Most animals have an internal or external skeleton to give them structure and so do not need cell walls.

Chloroplasts use light energy to make glucose and oxygen from carbon dioxide and water by photosynthesis. They are found only in some plant cells.

Chloroplasts – found only in some plant cells

Chloroplasts are structures that contain a chemical called chlorophyll which absorbs light energy to make food for the plant by a process called photosynthesis. Photosynthesis is actually a series of chemical reactions that are all controlled by enzymes.

Permanent vacuole – found only in plant cells

The vacuole in a plant cell contains water and some dissolved chemicals. This mixture is called cell sap. The vacuole acts a bit like a reservoir for the cell.

5. Copy and complete Table 5.1 to summarise the functions of the different cell parts.
6. A cell could be compared to a factory with the different parts working together so that the cell can function. Compare each part of a cell to part of a factory. For example:

   The mitochondria are like the generators in a factory because they release the energy from fuel.
7. The nucleus is sometimes called the 'brain' of a cell. Explain how much you agree with this statement.
8. People sometimes say they have a 'high metabolic rate' if they release energy from their food quickly. Use the information about the different cell parts to suggest why some people release energy from their food more quickly than others. Explain your answer.
9. Where in a plant would you expect to find cells with the most chloroplasts? Give a reason for your answer.

Cell part	Function	Found in...
Nucleus	Controls the chemical reactions in the cell	Animal and plant cells
Cytoplasm		
Cell membrane		
Mitochondria		
Ribosomes		
Cell wall		
Chloroplasts		
Vacuole		

Table 5.1

Most cells have a specific function to carry out. For example, nerve cells transmit electrical impulses around the body and cells in the pancreas produce insulin. Once these cells have specialised they keep producing these specialised cells. However, there are some cells, called stem cells, that can produce a range of different cells. Stem cells are found in embryos, adult bone marrow and some other body tissue such as liver tissue. Stem cells will be covered in Section 8.3.

How do dissolved substances move into and out of cells?

So far this chapter has referred to a range of processes, such as respiration and photosynthesis, that take place inside cells. These processes take place as a series of chemical reactions and therefore need chemicals that can get into a cell. They also produce chemicals that need to leave the cell. In this section you will consider how these substances move into or out of a cell.

In this chapter we are going to concentrate on the movement of oxygen and water into and out of cells. Both of these substances will, at times, enter or leave cells.

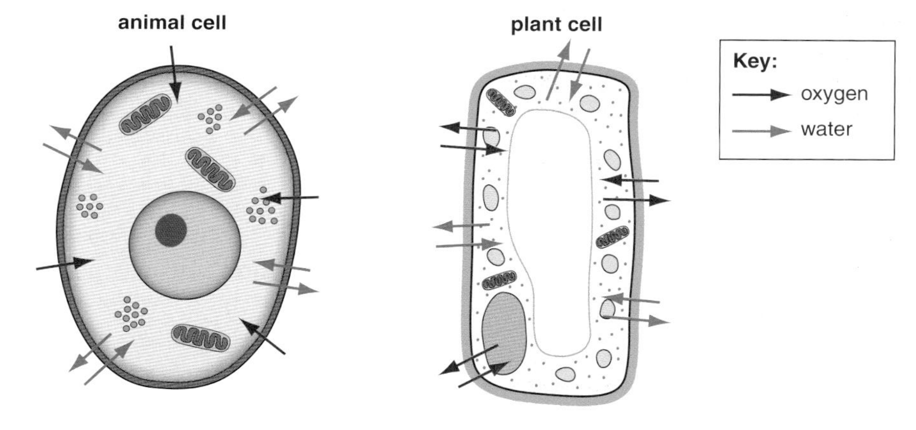

Figure 5.8 The movement of oxygen and water in and out of animal and plant cells

⑩ Make a list of substances that you think need to:
a) get into a cell;
b) leave a cell.

⑪ What do these substances need to pass through in order to enter or leave a cell?

Diffusion

Concentration is the amount of one substance in a given volume of another.

Diffusion is the movement of particles in a gas or any dissolved substance from a region of higher concentration to a region of lower concentration.

A **partially permeable membrane** is a membrane, such as a cell membrane, that will allow only some substances to pass through it.

If you spray air freshener in one corner of a room it is not long before it can be smelt throughout the room. This happens because the particles that make up the air freshener move around. To start with there are lots of particles in the space where you sprayed the air freshener, but quickly they spread out to where there are fewer particles. They move from a region where there is a higher **concentration** of particles to a region where there is a lower concentration. This process is called **diffusion** and it explains why gases spread out. Imagine if your class started huddled in the corner of the room and then ran about until they were evenly spread around the room. This gives a 'picture' or a model of how the particles behave when diffusion takes place.

Particles of a substance that is dissolved in a liquid will also diffuse. If you put a drop of ink in a glass of water you can see the ink gradually spread throughout the water. This happens because the particles in the ink are diffusing amongst the water particles. They are moving from where there is a higher concentration of 'ink' particles.

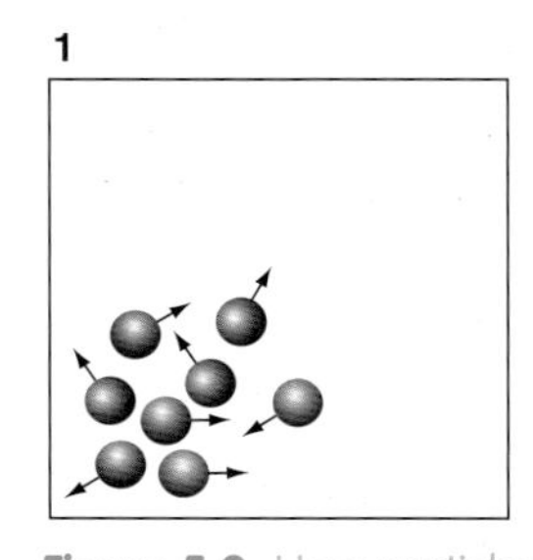

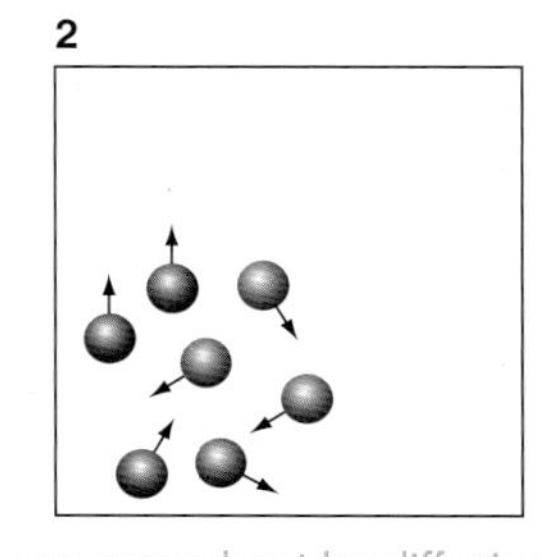

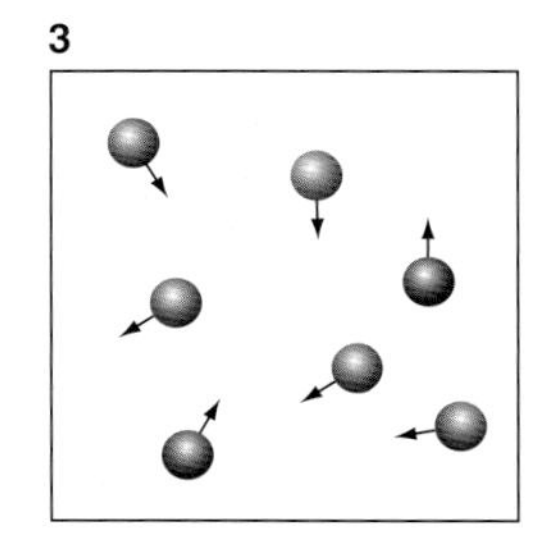

Figure 5.9 How particles in a gas spread out by diffusion

Oxygen will dissolve in water and so can diffuse when in solution. This is how it moves into cells when it is used for respiration. Since oxygen is used up quickly in the mitochondria for respiration, there will always be a lower concentration of oxygen inside the cell than in the liquid outside the cell. This ensures that the oxygen will always diffuse into the cell. The greater the difference in concentration, the faster the rate of diffusion.

Of course oxygen, or anything else trying to get into a cell, has to get through the cell membrane. Oxygen can do this by passing through the cell membrane. The membrane will not let through larger molecules such as starch. Therefore the cell membrane is referred to as a **partially permeable membrane**.

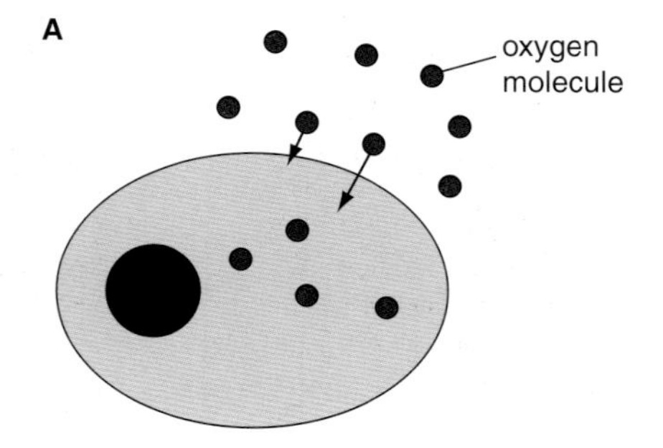

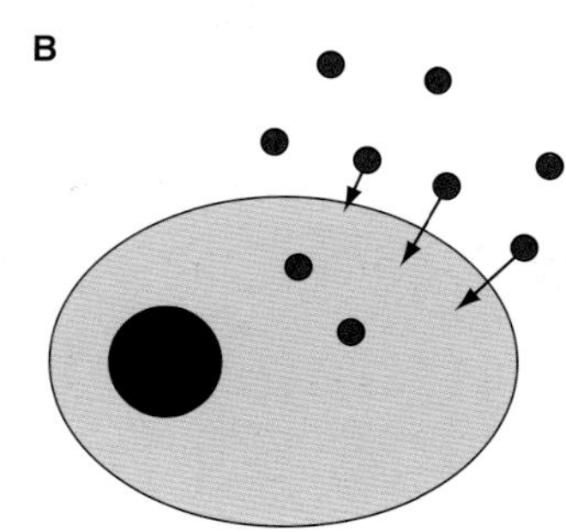

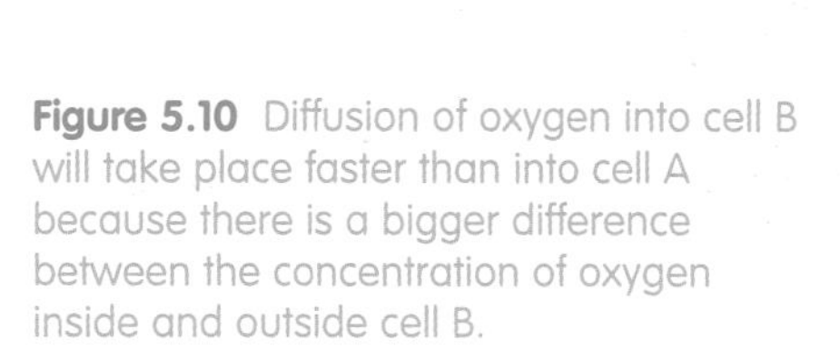

Figure 5.10 Diffusion of oxygen into cell B will take place faster than into cell A because there is a bigger difference between the concentration of oxygen inside and outside cell B.

⓬ If a slice of lemon is left in a glass of cranberry juice it turns red, even if it is taken out of the glass and rinsed under a tap. Explain why this happens. Try to use the words in the definition box, on p. 82, in your answer.

⓭ 'Stabilised liquid oxygen' (Figure 5.11) is a new product that claims to be able to increase your energy levels by raising the concentration of oxygen in the blood.

a) What effect would increasing the concentration of oxygen in the blood have on the rate of oxygen diffusion into a cell?

b) Explain your answer to part a). Use diagrams if they will help explain your ideas.

c) Explain how increasing the oxygen concentration of the blood could boost your energy levels.

d) There are a lot of people who do not agree that stabilised liquid oxygen actually works. What scientific ideas could they use to back up their point of view? Remember, this product is a dietary supplement.

Figure 5.11 Stabilised liquid oxygen – a new dietary supplement.

Osmosis

Having the correct amount of water in your blood is crucial, and this is controlled by your kidneys. You may remember learning about this in Chapter 1. If the concentration of water in your blood changes it will affect the cells in your body because water can diffuse into and out of cells. The diffusion of water into or out of cells is called **osmosis**. This is illustrated very clearly by red blood cells.

Red blood cells are specialised cells with a large surface area and no nucleus, a structure that makes them very good at carrying oxygen around the body. When they are in a solution with the correct concentration of water they have a recognisable 'bi-concave' shape. If the concentration of water around them is too low they shrivel up and if it is too high they swell up and can even burst. You can see a similar effect if you put a tomato in a glass of water and one in a glass of water with lots of salt dissolved in it.

Osmosis is the diffusion of water, through a partially permeable membrane, from a region of higher water concentration to a region of lower water concentration.

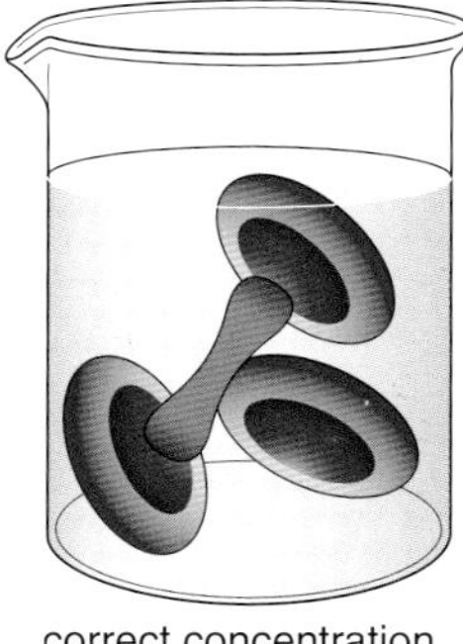

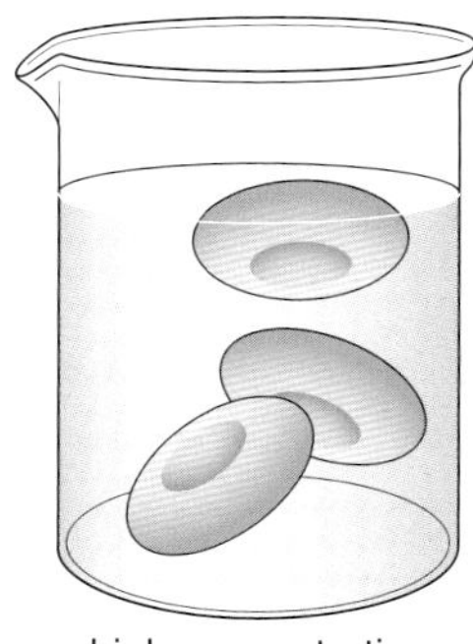

Figure 5.12 This is what happens to red blood cells in solutions with different concentrations of water.

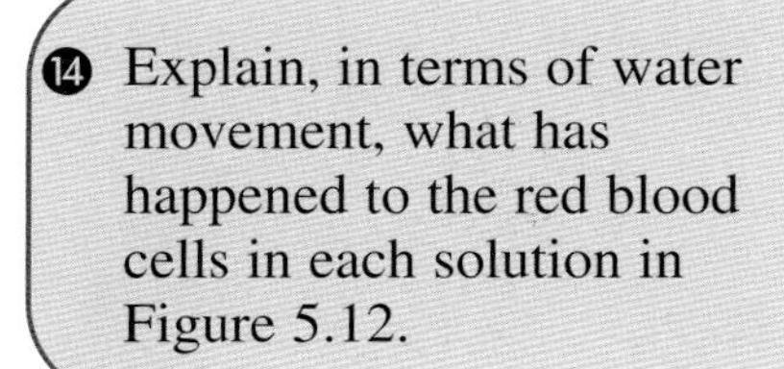

⓮ Explain, in terms of water movement, what has happened to the red blood cells in each solution in Figure 5.12.

What can affect osmosis?

To fully understand osmosis you have to consider the water molecules and any substances dissolved in the water inside and outside the cell. Look at Figure 5.13 and think about the number of water and glucose molecules in each solution.

In Figure 5.13 a) water moves from left to right, from a region of higher water concentration to a region of lower water concentration. However, some people will explain this by referring to the concentration of glucose in each solution. In this case you would say that water has moved from a less concentrated (dilute) glucose solution to a more concentrated glucose solution through a partially permeable membrane. In Figure 5.13 b) the water moves from right to left because the concentration of glucose is higher on the left than on the right. So, you can alter the movement of water by osmosis by adding more water or by adding more solute to the water.

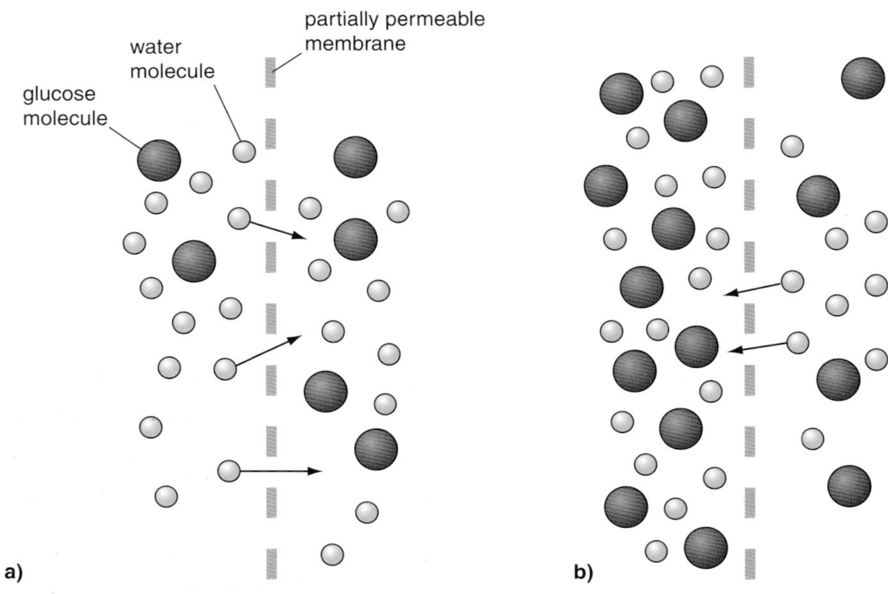

Figure 5.13 Two partially permeable membranes. Both membranes have the same concentration of solution on the right-hand side. However, water moves in opposite directions through the membranes.

15. Why is it important to avoid getting dehydrated when you are exercising?
16. Very occasionally marathon runners can in fact drink too much water and become over-hydrated. This is more likely to happen when they have been sweating heavily and have lost a lot of salt from their body.
    a) What effect will over-hydration have on the cells in the body?
    b) Why does losing a lot of salt from the body increase the effect of over-hydration on body cells?

17 For each diagram in Figure 5.14, state which way the water will move and explain why this will happen in terms of water and solute concentration.

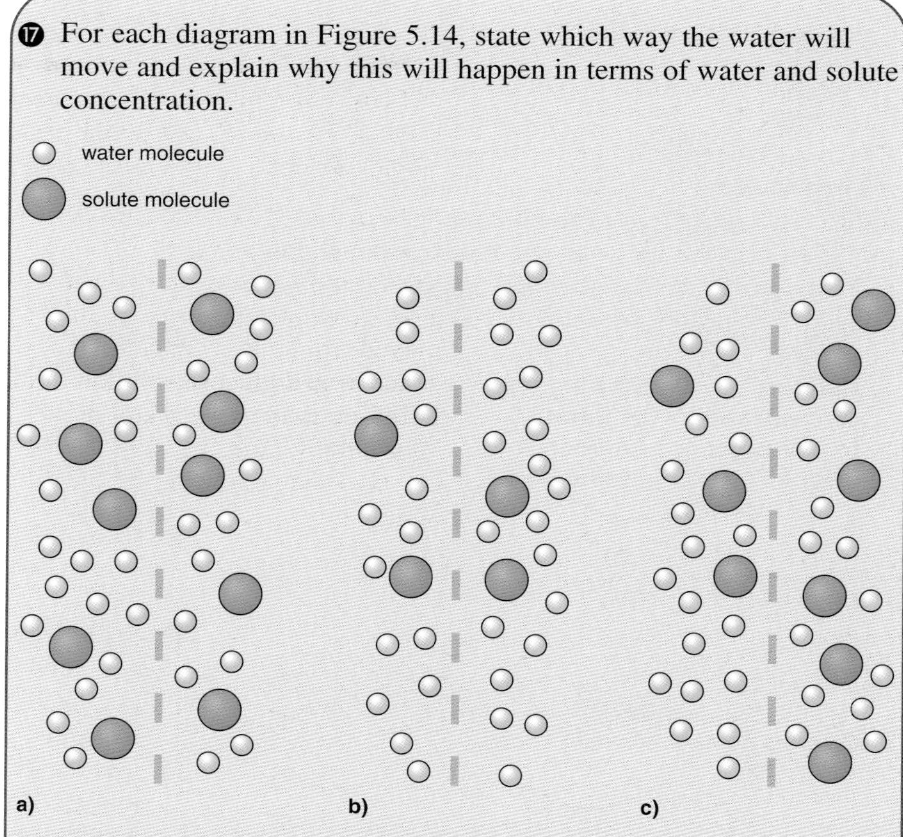

Figure 5.14 These partially permeable membranes have a different concentration of solution on each side.

18 Sports drinks can be divided into three types. Hypotonic drinks have a sugar concentration lower than your body fluid, isotonic drinks have the same concentration of sugar as your body fluid and hypertonic drinks have a higher sugar concentration than your body fluid.

a) Which type of sports drink would be best to drink if you wanted to re-hydrate your cells quickly?
b) Explain your answer to part a).
c) What are the benefits and drawbacks of drinking hypertonic drinks if you go for a long bike ride on a hot day?

Activity – Changing the rate of osmosis

Two students carried out an investigation using cylinders of potato. They cut out five potato cylinders and measured the mass of each one. They made up five glucose solutions with a different concentration and left a potato cylinder in each solution for 24 hours. After 24 hours they dried each potato cylinder and measured its mass again. The results are shown in Table 5.2.

Glucose concentration in moles per litre	Starting mass of potato cylinder in grams	Mass of potato after 24 hours in grams
0	7.76	9.38
0.2	8.01	8.04
0.4	7.96	7.21
0.6	7.92	7.78
0.8	7.93	6.65

Table 5.2 The results from the five potato cylinders left in different glucose solutions for 24 hours

1. Why did the students dry each potato cylinder before measuring its mass after each experiment?
2. What possible variables should be kept the same to make this investigation a fair test?
3. Work out the change in mass of each cylinder. These mass changes will be positive for cylinders that got heavier and negative for those that got lighter.
4. Now plot a graph for this data showing glucose concentration on the *x*-axis and change in mass on the *y*-axis. Think carefully about how you set this out so that you can plot positive and negative values on the same graph.
5. One of the results appears to be anomalous. Which one is it and what may have caused this?
6. Write a conclusion for this investigation. You will need to describe the pattern shown by the graph and use scientific ideas to explain what happened.
7. Use the graph to predict the concentration of glucose that you would use that would cause no change in mass of the potato. Explain, in terms of osmosis, why this would happen.

Summary

- ✓ All living things are made up of cells.
- ✓ Most animal cells have a nucleus, cytoplasm, a cell membrane, **mitochondria** and **ribosomes**.
- ✓ In addition to these parts, plant cells often have **chloroplasts** and a permanent vacuole.
- ✓ Each cell part has a specific function.
- ✓ The chemical reactions that take place inside a cell are controlled by enzymes.
- ✓ Some cells are specialised to carry out a particular function.
- ✓ The structure of a specialised cell relates to its particular function.
- ✓ **Diffusion** is the spreading out of particles of a gas or a substance dissolved in a liquid.
- ✓ Particles move by diffusion from a region of higher **concentration** to a region where there is a lower concentration.
- ✓ Oxygen diffuses into and out of cells through a **partially permeable membrane**.
- ✓ Water moves through a partially permeable membrane by **osmosis** from dilute solutions to more concentrated solutions.

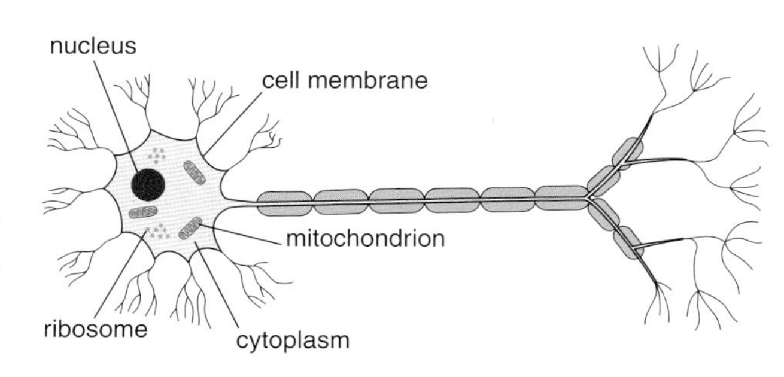

Figure 5.15 A nerve cell with some cell parts labelled

1. a) Name the labelled part of the cell in Figure 5.15 that releases energy from glucose. *(1 mark)*
   b) Name the part of the cell that synthesises enzymes. *(1 mark)*
   c) A nerve cell is a specialised cell that carries electrical nerve impulses throughout the body.
      i) Describe one feature of a nerve cell that allows it to carry out this function. *(1 mark)*
      ii) Explain how this feature allows a nerve cell to carry out this function. *(1 mark)*

2. a) Red blood cells (Figure 5.16) are specialised to carry oxygen around the body. Explain how the structure of a red blood cell relates to its function. *(2 marks)*

Figure 5.16 A red blood cell

   b) Explain how oxygen enters a red blood cell. *(3 marks)*
   c) When mountaineers climb high mountains the concentration of oxygen in the air they are breathing decreases. Explain how this will affect the uptake of oxygen by their red blood cells. *(2 marks)*

3. A chip shop owner claimed that if he soaked his potato chips in water before he fried them they were firmer than if they were cut up and fried without soaking them.
   a) Explain how scientific ideas can be used to back up his claim. Use correct scientific words in your explanation. *(4 marks)*
   When some uncooked chips were soaking, a worker in the shop poured some salt into the water and left the chips soaking in salty water.
   b) Suggest how this may have affected the potato chips. *(2 marks)*
   A student carried out an experiment to see if adding salt to the water that potato chips were soaking in affected the amount they changed in mass while soaking. She soaked one potato chip in pure water and one in a concentrated salt solution. She measured their masses before and after they had been soaked.
   c) i) State the dependent variable in this investigation. *(1 mark)*
      ii) State the independent variable in the investigation. *(1 mark)*
   d) Do you think that the results from this investigation would provide enough information to draw a firm conclusion about the effect of salt concentration on the change in mass of potato chips? Explain the reasons for your answer. *(3 marks)*

Chapter 6
What are the energy inputs and energy losses in a food chain?

At the end of this chapter you should:

- ✓ understand the process of photosynthesis;
- ✓ know the factors that affect the rate of photosynthesis;
- ✓ know how plants use the products of photosynthesis;
- ✓ be able to apply knowledge of the factors affecting photosynthesis in a range of habitats and agricultural situations;
- ✓ understand why plants need minerals and the effects of a deficiency;
- ✓ understand the construction of pyramids of biomass;
- ✓ be able to apply biomass and energy principles to the efficiency of food production;
- ✓ be able to evaluate the conflicts and compromises between feeding the population and damaging the environment;
- ✓ understand the role of microorganisms in decomposition and recycling carbon and plant nutrients.

Figure 6.1 The start of the food chain in biodiverse rough pasture. The inset shows the stomata through which the carbon dioxide diffuses into the leaf for photosynthesis.

Photosynthesis

The word **photosynthesis** can be split into two parts, 'photo' meaning 'light' and 'synthesis' meaning 'to make'.

Green plants use light to convert the simple molecules carbon dioxide and water into larger glucose molecules using enzymes in chloroplasts. This process is called photosynthesis.

We can break down the summary equation for photosynthesis into three parts, as shown below:

Raw materials	Conditions	Products
carbon dioxide + water $6CO_2 + 6H_2O$	light $\longrightarrow$ enzymes chlorophyll	glucose + oxygen $C_6H_{12}O_6 + 6O_6$

How is the light captured?

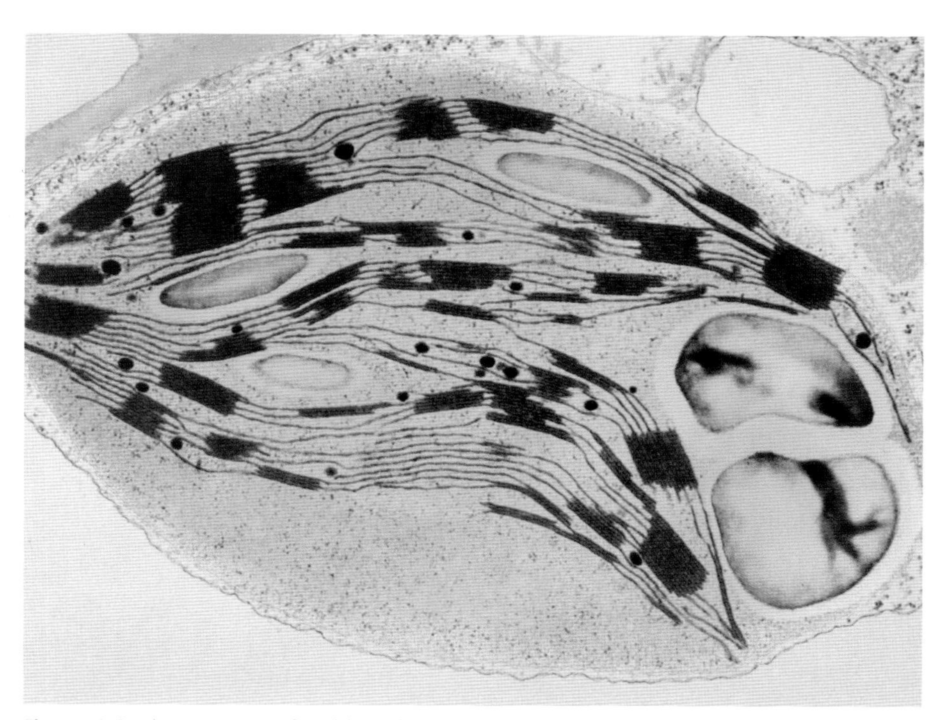

Figure 6.2 The structure of a chloroplast (magnification about 50,000)

You studied plant cell structure in Section 5.1 and will recall that some plant cells have chloroplasts to absorb light energy. Figure 6.2 shows that the chloroplast has many membranes inside it. These membranes contain chlorophyll, which makes the leaves appear green. When light reaches the chlorophyll, enzymes in the chloroplasts use this energy to catalyse reactions that produce glucose. Glucose is quickly converted to starch that can be stored in the leaf, and oxygen is released as a waste product.

How is the raw material, carbon dioxide, obtained?

The carbon dioxide concentration of the atmosphere can be written as a percentage or as ppm (parts per million). Atmospheric carbon dioxide levels are changing, and in 2006 a record high level of 381ppm was recorded.

Stomata are the tiny openings in the surface of a leaf through which gases can pass by diffusion.

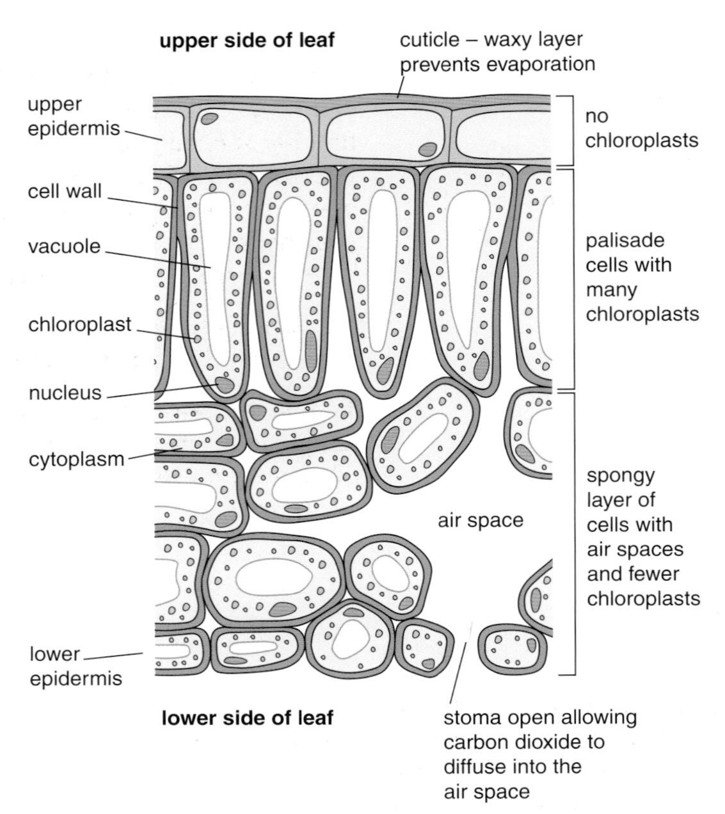

Figure 6.3 A section of a leaf showing the open stomata (singular 'stoma') on the lower surface

1. What materials are needed for photosynthesis?
2. What is the energy input for photosynthesis?
3. A variegated plant has many white leaves.
   a) What essential condition for photosynthesis is missing from these leaves?
   b) Explain why this plant would not grow as rapidly as a normal green-leaved plant.
4. What are the products of photosynthesis?

During the day, the leaf is photosynthesising and using carbon dioxide. The **stomata** (small pores) are open and the carbon dioxide diffuses into the leaf from the air. Diffusion always takes place from a region of higher concentration to a lower concentration. Although the concentration of carbon dioxide in the air is on average only about 0.037% (or 370 ppm), when the leaf is photosynthesising it will be using up the carbon dioxide inside the leaf and so maintaining the concentration difference needed for diffusion.

How is the raw material, water, obtained?

Roots take up water from the soil by osmosis (see Section 5.2). The water is then transported up the stem to the leaves. Water is essential for support of the leaf cells; if there is a shortage of water, the stomata will close to reduce water loss and photosynthesis will stop. This may happen around midday on a hot sunny day when there is a high water loss from the leaves, or during drought conditions when the soil is very dry.

What factors affect the rate of photosynthesis?

The **rate of photosynthesis** is the speed at which the photosynthesis reaction takes place. It can be measured as the rate of formation of oxygen.

A **reaction rate** is the quantity of product produced or reactant used per unit time. You might compare this to a model with which you are familiar such as the money you earn for washing cars on a Saturday job. For example, if you get paid £2 per wash and you wash two cars an hour, your hourly rate is £4.00 per hour. If you wash five cars an hour, your hourly rate is £10 per hour. The more cars you can wash in an hour, the greater your hourly rate of pay.

You can measure the rate of photosynthesis by finding the volume of oxygen produced per hour using a range of input values.

To find out how the input of carbon dioxide, light intensity and temperature limit the rate of photosynthesis, we need to measure the rate of photosynthesis.

Activity – Analysing the effect of carbon dioxide concentration on the rate of photosynthesis

A group of students decided to investigate the effect of carbon dioxide concentration on the rate of photosynthesis. They decided to calculate the rate of photosynthesis by measuring the volume of oxygen produced from an aquatic plant, *Elodea* (see Figure 6.4).

To investigate the effect of carbon dioxide concentration, the students made a **series dilution** of sodium hydrogencarbonate solution. They started with a 5.0% solution. They followed these instructions to make solutions with different concentrations.

A **series dilution** is a range of concentrations. A minimum of five dilutions from concentrated to weak would give the input variable for the carbon dioxide concentration. The students planned the dilutions to give sensibly spaced points on the *x*-axis of their graph.

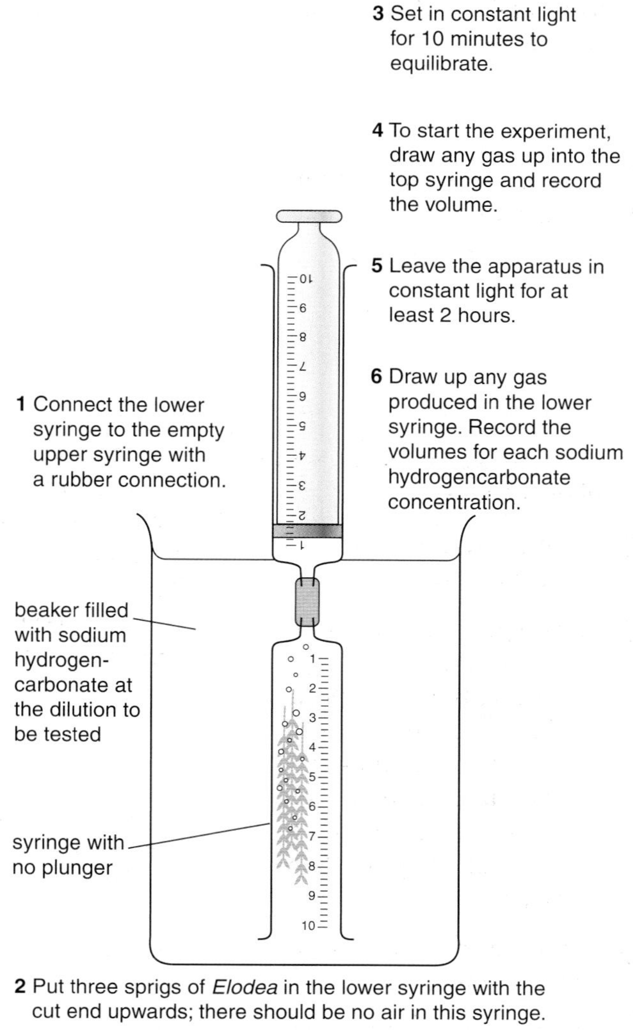

1 Connect the lower syringe to the empty upper syringe with a rubber connection.

2 Put three sprigs of *Elodea* in the lower syringe with the cut end upwards; there should be no air in this syringe.

3 Set in constant light for 10 minutes to equilibrate.

4 To start the experiment, draw any gas up into the top syringe and record the volume.

5 Leave the apparatus in constant light for at least 2 hours.

6 Draw up any gas produced in the lower syringe. Record the volumes for each sodium hydrogencarbonate concentration.

beaker filled with sodium hydrogen-carbonate at the dilution to be tested

syringe with no plunger

Figure 6.4 The apparatus and method used to measure rate of photosynthesis

1 Take six 1000 cm^3 beakers. Label them A to F.
2 Fill beaker A with 5% solution.
3 Use a measuring cylinder to add 900 cm^3 of water into beaker B.
4 Using a second measuring cylinder transfer 100 cm^3 sodium hydrogencarbonate solution from A to beaker B.
5 Stir beaker B with a glass rod to mix. This gives a 0.5% sodium hydrogencarbonate solution.
6 Use the 0.5% solution to make up the solutions in the proportions shown in Table 6.1.

❶ Copy and complete the final column of Table 6.1.

Beaker	Starting concentration of $NaHCO_3$	Volume of $NaHCO_3$ in cm^3	Volume of H_2O in cm^3	Final concentration
B	5%	100	900	0.5%
C	0.5%	800	200	
D	0.5%	600	400	
E	0.5%	400	600	
F	0.5%	200	800	

Table 6.1 A summary of the series dilution

The students used the apparatus and method shown in Figure 6.4 to investigate rates of photosynthesis. The students knew that they should only change one input variable in each investigation. The input variable was the range of concentrations of sodium hydrogencarbonate solution. They knew that the light intensity and temperature should be kept constant.

❷ What factor were the students controlling in this experiment?
❸ This experiment was completed outside on a hot June day. What factors may not have been controlled?

To develop the experiment further the students decided to calculate the rate of photosynthesis in each experiment as the rate per gram dry mass per hour. They measured the total dry mass of the *Elodea* used each time by placing it in a folded filter paper in a drying cupboard set at 100 °C. The samples were dried and weighed every hour until the mass of dried *Elodea* remained constant for two successive weighings.

Table 6.2 shows the set of results they obtained.

4. Copy the table and complete the calculations. Use the following equation to complete the final column:

$$\frac{\text{volume of oxygen in cm}^3}{\text{mass of } Elodea \text{ in g}} \times \frac{1}{\text{time (h)}}$$

5. Use the results in the table to construct a graph with volume of oxygen in cm^3/g/h on the *y*-axis and concentration of sodium hydrogencarbonate solution on the *x*-axis.

6. Suggest a suitable title for this graph, which links the output variable with the input variable.
7. Describe the pattern shown by your graph.
8. Suggest how you would modify this experiment to investigate the effect of light intensity on the rate of photosynthesis.

Concentration of $NaHCO_3$ as %	Starting volume in cm^3	Final volume in cm^3	Volume of oxygen in cm^3	Mass of *Elodea* in g	Time in h	Volume of oxygen in cm^3/g/h
0.5	0.2	9.6	9.4	1.341	2	
0.4	0.4	8.6	8.2	1.399	2	
0.3	0.3	7.1	6.8	1.439	2	
0.2	0.5	5.9	5.4	1.510	2	
0.1	0.4	4.8	4.4	1.522	2	

Table 6.2

A **limiting factor** is the one that controls the rate of the product formation.

From the equation for photosynthesis given in Section 6.1 you can see that carbon dioxide is one of the raw materials. Although leaf cells respire 24 hours of the day, carbon dioxide cannot leave the leaf at night, since the stomata are closed (light intensity controls when the stomata open). By daybreak there will be a build-up of carbon dioxide in the leaf from the respiration which has taken place overnight. Initially light intensity will be a **limiting factor** for photosynthesis, but as the day gets brighter there will be more rapid photosynthesis, using up the carbon dioxide.

Carbon dioxide forms a small percentage of atmospheric gases and can be a limiting factor when the light intensity is high.

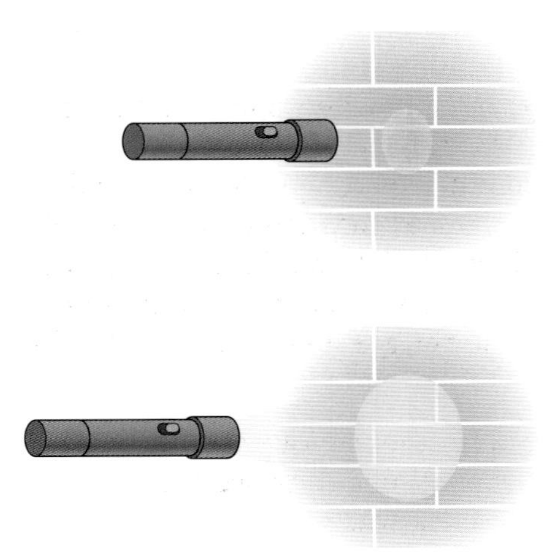

Figure 6.5 The light from the torch remains the same. As the torch get further away the light spreads over a larger area, so the light (energy) per unit area is less.

Light intensity is the amount of light (energy) per unit area of the surface. Light intensity decreases as the square of the distance from the light source.

How does light intensity affect the rate of photosynthesis?

First, work through this simple model to make sure you understand what light intensity means. If you switch on a torch, hold it close to a wall and shine the beam at the wall, you get a small circle of light. As you move the torch further from the wall the circle of light gets larger and dimmer. The light energy given out by the torch is the same, but the light energy falling on each square centimetre of the wall (the **light intensity**) is less when the torch is further away.

The decrease in light intensity is proportional to $1/d^2$, where d is the distance from the light source. This relationship shows that as the distance from a light source is increased, the light intensity reduces very rapidly (see Figure 6.6). You could use this graph of light intensity against distance if you were to plan an experiment to find how light intensity affected the rate of photosynthesis. Unfortunately, normal light bulbs are very inefficient and give off heat as well as light. Could this introduce another variable when the lamp is very close to the apparatus? Energy-efficient fluorescent bulbs give off less heat.

Light intensity generally increases towards midday and then decreases towards evening, but can change quite quickly on a cloudy day. Light is the energy input and will have a significant effect on the rate of photosynthesis. The light (energy) capture stage of photosynthesis depends on the light intensity and is independent of temperature.

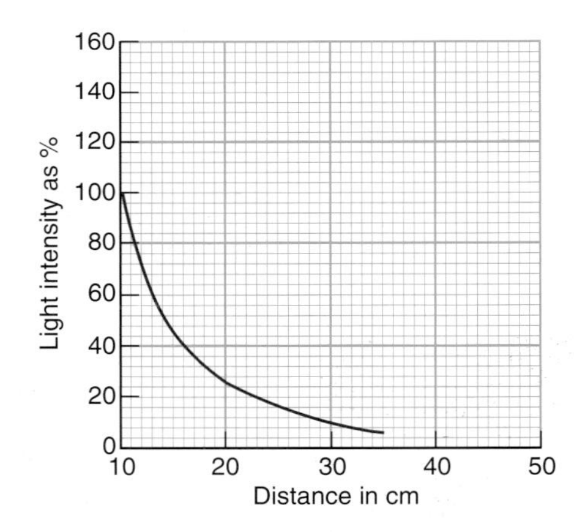

Figure 6.6 How light intensity decreases with increase in distance. For this graph the light intensity was taken as 100% at 10 cm distance.

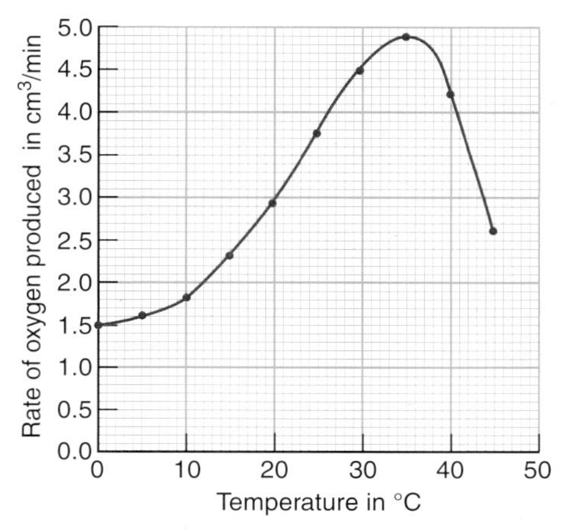

Figure 6.7 Effect of temperature on the rate of photosynthesis at a fixed light intensity and fixed carbon dioxide concentration.

How does temperature affect the rate of photosynthesis?

After the light has been captured, enzymes in the chloroplasts catalyse the chemical reactions that convert carbon dioxide to glucose. Therefore, you would predict that the rate of glucose production would increase with increase in temperature up to the optimum temperature of the enzymes (enzymes are covered in Sections 7.1 and 7.2). The rate of photosynthesis increases on a warm day since the enzyme reactions are temperature-dependent. The production of glucose and hence starch will be slower on a cold day.

Figure 6.7 shows that at higher temperatures the rate of photosynthesis is faster, for a fixed concentration of carbon dioxide. Above the optimum temperature, photosynthesis slows down.

Having worked through this section, you can see that the rate of photosynthesis at any time is a result of the interaction of several factors. Any one factor of low temperature, low light intensity or low levels of carbon dioxide could limit the rate of photosynthesis.

6.3 How does the plant use glucose, the end product of photosynthesis?

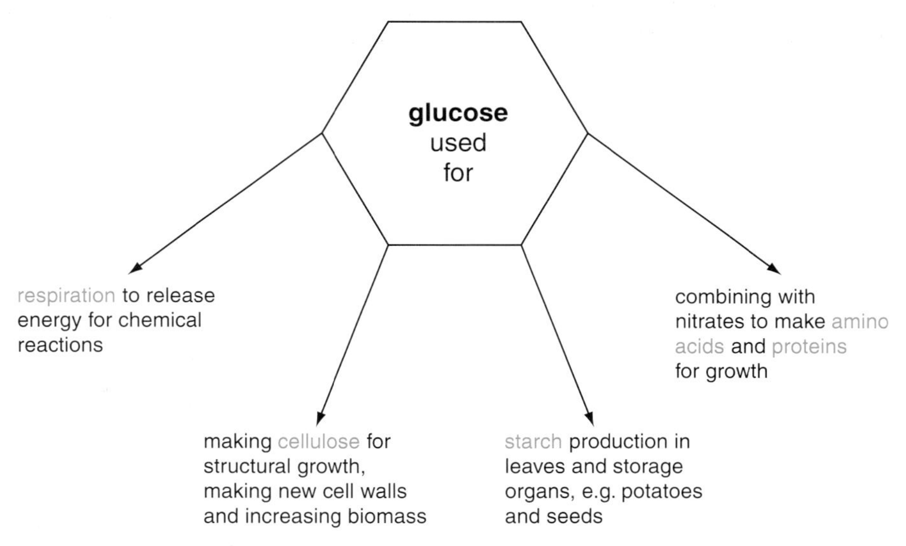

Figure 6.8 How the plant makes use of the glucose produced

Figure 6.9 A shortage of magnesium ions makes a plant's leaves turn yellow.

After glucose is produced, it is converted to starch in the leaf. We can test for starch to find out whether a leaf has been photosynthesising (Figure 6.10). Materials can only be carried to other parts of the plant when dissolved in water, so the insoluble starch is converted to soluble sugar for transport throughout the plant.

All parts of the plant need a supply of glucose for respiration. The growing shoots of the plant need glucose for respiration, to release chemical energy for making new cells. Just as you need minerals to keep healthy, so do plants. Nitrates are required for making amino acids and therefore for building proteins to make the cytoplasm and enzymes in the new cells. Magnesium ions are needed to produce chlorophyll molecules. The mineral ions are taken up from the soil into the roots. If plants are unable to produce sufficient protein, they will grow very slowly and remain small. If they are unable to obtain enough magnesium they will not be able to make chlorophyll, so their lower leaves will become yellow and not photosynthesise efficiently.

1 Dip
Take a leaf from a plant standing in sunlight and dip in boiling water for 10 seconds to denature enzymes and make the membrane permeable.

THEN TURN OFF THE BUNSEN because alcohol vapour is inflammable.

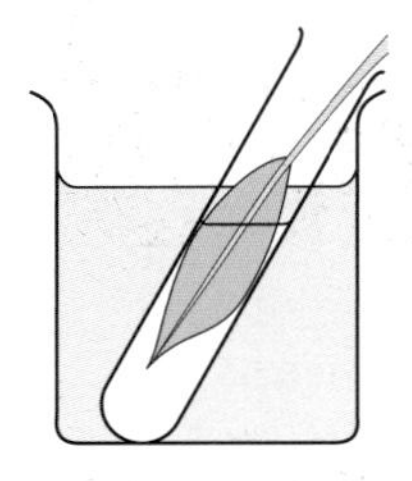

2 Decolorise
Put tube of alcohol into the water bath and allow to warm up. Add leaf and shake gently until alcohol becomes green.

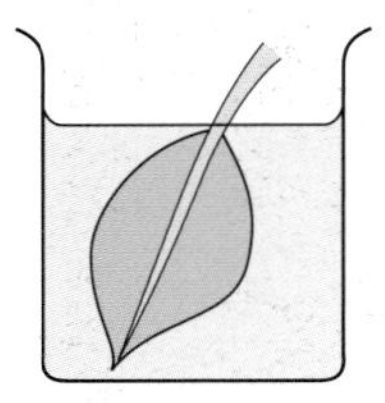

3 Soften
Soften the leaf by rinsing in the hot water.

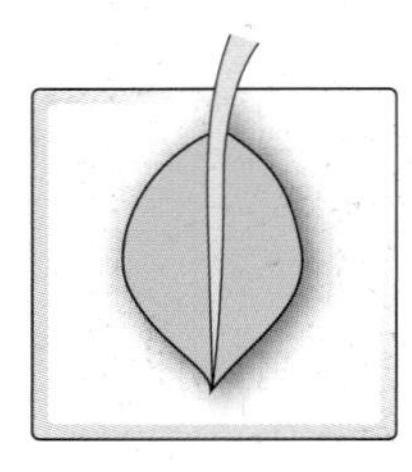

4 Flatten
Flatten the leaf underside up on a white tile and add a few drops of iodine. Where starch is present it will give a blue black colour.

This method can be used with a leaf which has been partly covered with a stencil or photonegative, or with a variegated leaf.

Figure 6.10 Testing a leaf for starch

5. A plant was put into a black plastic bag and left for 24 hours. The leaves were then tested for starch. Explain why you would not expect to find starch in the leaves even during the day.
6. Where are the enzymes located that control the conversion of carbon dioxide to glucose?
7. When testing a leaf for starch it is dipped into boiling water to stop enzyme reactions.
   a) Suggest which enzyme reactions these are.
   b) Why does the alcohol become green?
   c) If parts of the leaf become blue-black when tested with iodine solution, it indicates that starch is present. What could you conclude if the iodine solution stayed brown?
8. Draw sketch graphs to compare the rate of oxygen produced on a bright winter day and on a bright summer day. Assume the light intensity is the same on both days.
9. A student grew a tomato plant in sand. After a month he compared it with his friend's that had been grown in potting compost. Both plants had been watered regularly but the plant in sand was shorter, had much smaller leaves and the lower leaves were yellow. Suggest reasons for these differences.

What factors limit the rate of photosynthesis in the environment?

We have seen that under laboratory conditions light intensity, carbon dioxide concentration and temperature can all affect the rate of photosynthesis.

Activity – From laboratory to outdoors

Use the information gained from laboratory examples to consider photosynthesis in the different environments shown below.

A field crop on a hot, sunny day

Figure 6.11

1. Will light intensity be a limiting factor?
2. By 9 am the rate of photosynthesis has reached a constant level and does not increase even though the light intensity is increasing. What could be the limiting factor?
3. At night the stomata are closed and the plant continues to respire resulting in a build up of carbon dioxide inside the leaf. What effect would you expect this to have on the rate of photosynthesis when it first gets light in the morning?

A forest in Iceland

It has been said that you can't get lost in a forest in Iceland because the birch trees only grow to about 1.5 m high (see Figure 6.12). Iceland has periods of 24 hours darkness in winter and 24 hours light in summer, although the light intensity is not the same during those 24 hours.

4. What factors limit the growth of the trees?
5. Is carbon dioxide concentration likely to be a limiting factor?

Figure 6.12

A pond

Figure 6.13

Geese use the pond and their waste has resulted in extra nitrates (plant food) in the water. As a result there is dense weed growth and many snails feeding on the weeds.

6. What do you think will happen to the carbon dioxide concentration in the water overnight? Explain your answer.
7. During mid-morning streams of tiny bubbles can be seen rising to the surface. What gas do you think these bubbles contain, and where does it come from?
8. What do you think happens to the carbon dioxide content of the water during the day? Explain your answer.

Can an understanding of photosynthesis be used to increase greenhouse crop production?

For maximum profit, greenhouse crops must have a high yield and good quality. Producers also need to raise the crops to maturity in the minimum possible time. You saw in the activity in Section 6.2 that carbon dioxide could be a limiting factor for photosynthesis. Many experiments have been carried out to find the optimum light intensity and carbon dioxide concentrations for different plants.

Figure 6.14 A large commercial greenhouse

Activity – Greenhouse crop production

Use the graph in Figure 6.16 to answer the following questions.

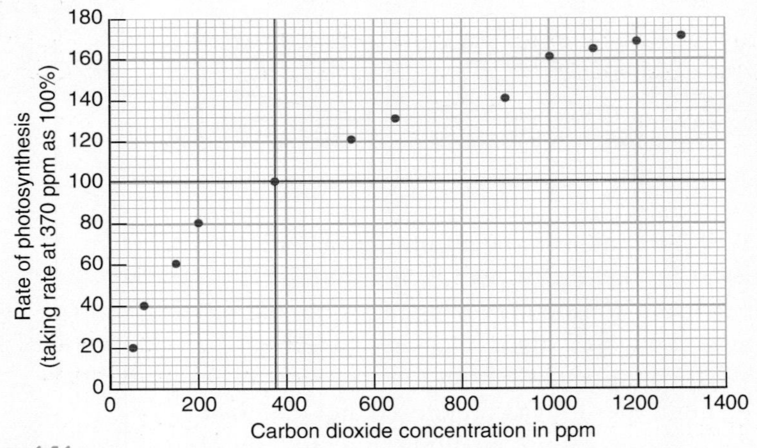

Figure 6.16

Figure 6.15 A computer control panel that controls conditions within a greenhouse

In a closed greenhouse in winter with artificial lighting and no ventilation, the levels of carbon dioxide can drop during the day to 200 ppm. (The rate of photosynthesis at 370 ppm carbon dioxide (atmospheric level) is taken as 100%. This is shown in red on the graph in Figure 6.16.)

1. Use the graph to find the effect on the rate of photosynthesis of reducing carbon dioxide levels from 370 ppm to 200 ppm.

Increased carbon dioxide concentration results in larger leaves, which increases the size and quality of lettuces. Colorado State University found that increasing carbon dioxide levels to 550 ppm meant that they could only fit 16 instead of 22 lettuces into a box.

2. Use your knowledge of photosynthesis to explain why the lettuces were bigger.

❸ From the graph, by what percentage was the rate of photosynthesis increased by changing from 370 to 550 ppm carbon dioxide?

If the carbon dioxide concentration is raised to 1000 ppm, for salad crops such as tomatoes, peppers and cucumbers, they flower earlier producing more flowers and more fruits.

❹ What shape is the curve above a concentration of 1000 ppm?

❺ Suggest why the grower does not usually increase the concentration above 1000 ppm.

Victorian greenhouses were heated by manure. The decomposition produced sufficient heat for fruits such as peaches to be produced early in the season. The rotting manure also raised carbon dioxide levels.

In modern greenhouses the carbon dioxide concentration can be raised by using a heater burning oil. This will increase the temperature as well as carbon dioxide levels. However, there are problems:

- the water produced during combustion increases the growth of mildew;
- if impure fuel is burnt, sulfur dioxide and oxides of nitrogen are given off; these cause acid rain-type damage to leaves, thus reducing photosynthesis.

The best, but not the cheapest, way of supplementing the carbon dioxide is to use storage cylinders of liquid carbon dioxide and computer control to regulate carbon dioxide gas concentration within the greenhouse.

❻ Sensors linked to a computer measure the carbon dioxide concentration. Suggest why the computer is programmed to activate a valve that increases the carbon dioxide concentration as the light intensity increases.

Pyramids of biomass

In Section 6.3 you saw that the products of photosynthesis can be used to increase plant **biomass**.

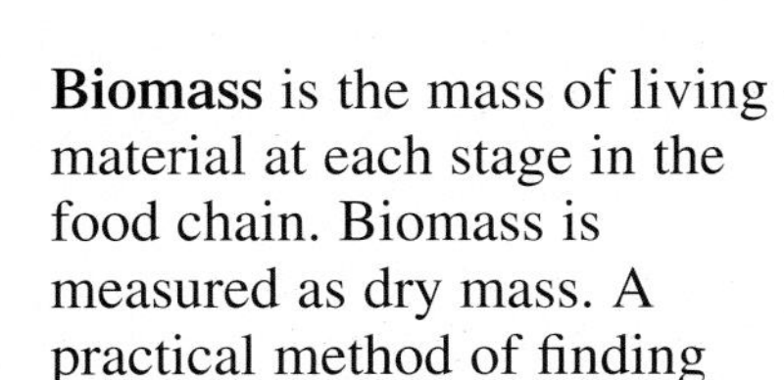

Biomass is the mass of living material at each stage in the food chain. Biomass is measured as dry mass. A practical method of finding dry mass was given in the activity in Section 6.2.

To recap, remember that:

- light (energy) is trapped by the chlorophyll;
- this energy is used to 'fix' carbon dioxide into a carbohydrate form (glucose); and
- carbohydrate is converted in the plant to cellulose cell walls, plant protein or stored as starch.

Plants (primary producers) are at the start of the food chain and are eaten by herbivores (primary consumers). The plant material eaten is used by the herbivore for respiration, in order to release the energy needed for maintaining body temperature, for movement and for growth of new tissues. The herbivore can be eaten by a carnivore, and once again some energy is used for movement and maintaining body temperature and some energy is transferred to new carnivore tissue.

To draw a **pyramid of biomass** (Figure 6.17) you need to:

- sample all the organisms to estimate the size of the population;
- examine the size distribution of organisms and find the dry mass of a representative sample;

A **pyramid of biomass** is a diagram that illustrates quantitatively the change in mass of living material at each level of a food chain.

A **trophic level** is a feeding level in a food chain. There can be more than one organism at each trophic level.

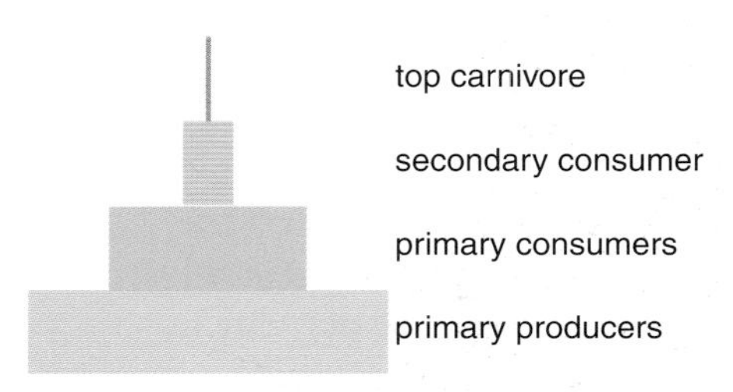

Figure 6.17 A pyramid of biomass

- calculate the biomass of each species;
- group the organisms into **trophic** (feeding) **levels** in a food web and calculate the biomass for each level;
- construct a pyramid on graph paper with the area of each rectangle proportional to the biomass.

Gaining all this information is not easy. There are problems:

- Obtaining dry mass of an animal means killing it, and you would certainly have problems obtaining dry mass if your herbivore were a cow! Hence, we use calculated data.
- The pyramid shows the biomass of what is growing at that moment. A tree for instance may have accumulated its biomass over 30 years, in contrast with green algae that grow quickly, reproduce and die within a few weeks.
- There are seasonal variations with some plants growing quickly in spring, seeding and dying by mid-summer.
- Anyone who mows a lawn knows that the productivity of a regularly mown lawn is greater than one that is neglected to seed in summer. This also applies to grazed meadows.

If you construct a pyramid of biomass to scale you will usually see a reduction in the biomass at each trophic level. If you examine the organisms in the food web used to construct the pyramid, you will see that there are large numbers of small organisms at the lowest feeding level and that the numbers decrease, but the size of organism increases, as you move up the feeding levels.

Pyramids of energy

A **pyramid of energy** is a diagram that illustrates quantitatively the decrease in the energy contained in biomass in each level of a food chain.

Efficiency of energy transfer is the percentage of the energy that is passed on to the next organism in the chain and is not lost.

For a **pyramid of energy**, if there is less biomass at each successive level there must also be less energy contained in the biomass. It is easier to examine energy loss in a food chain for animals. At each link, energy is released from the food by respiration and used for the chemical reactions taking place in the cytoplasm, for nerve impulse conduction, for muscle contraction, for movement and for keeping the body temperature constant and above that of the environment. When the body temperature is higher than the environment, energy will be lost as heat. Chemical energy is lost from the food chain in faeces and urine. This energy passes to the decomposers, which are not part of the food chain.

Taking the light input as 1000 units, the food chain in Figure 6.19 shows how little of that energy input reaches the secondary consumer, the fox. However, the percentage energy transfer is different in each ecosystem.

Can the biomass and energy principles be applied to food production?

Food production must be managed if the population is to be fed. Reducing the number of links in the food chain increases the efficiency

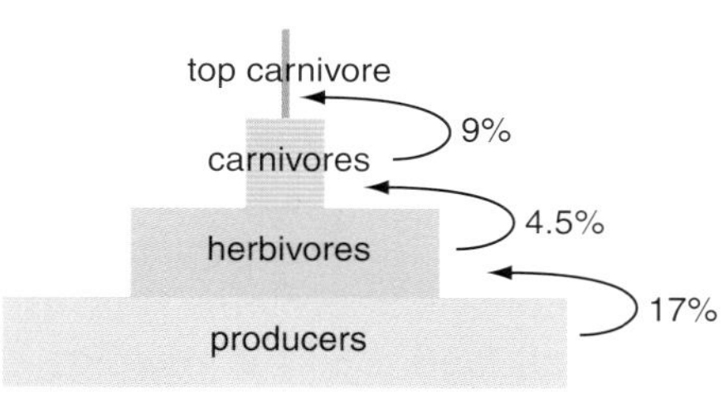

Figure 6.18 A pyramid of energy shows the **efficiency of energy transfer** from one trophic level to the next.

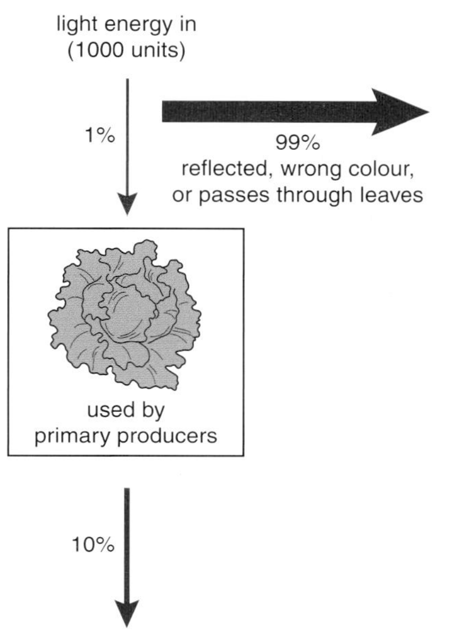

Figure 6.19 Food chain for a fox, showing energy losses at each stage

of food production, so that a greater amount of food is available to feed more consumers. By examining the pyramid of energy in Figure 6.18 we can see that if humans are the secondary consumer (meat eater) they receive less of the energy input than if they were the primary consumer (eating as a herbivore).

By looking at the energy conversion for a cow (Figure 6.20) you can see just how little energy is converted to increased biomass (muscle tissue) over the period of time that the animal is growing, typically 30 months.

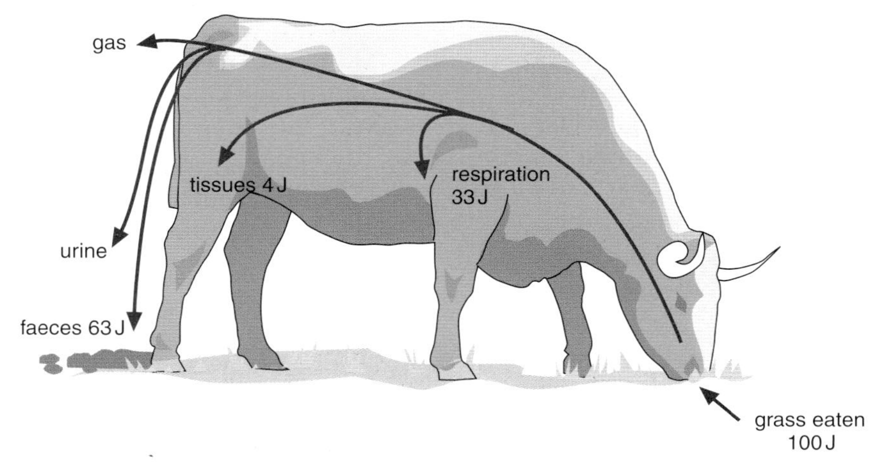

Figure 6.20 For every 100 J of energy available in the grass, only 4 J get into the animal's tissues

This gives us some points to think about.

- Would less energy be lost in producing animals for human consumption with a shorter 'age to slaughter', such as rabbits, chickens or ostriches?
- Could we reduce the amount of energy used by beef cattle?
- Would it be more energy-efficient for humans to consume more at the herbivore level (i.e. eat more vegetarian food)?
- Should we use biotechnology to produce more non-animal foods such as mycoprotein, frequently marketed as Quorn™?

The answer to all these questions is probably 'yes,' but in each case, traditions in society can result in a pre-determined pattern of behaviour. Scientific evidence is not generally taken into account when making decisions about what to eat. Here are some examples of human responses that you could use as discussion points.

- Beef is a traditional food and people would be slow to convert to eating rabbit meat. How quickly are people likely to change their established patterns of eating?
- The energy used by beef cattle could be reduced by restricting their movement and keeping them in an insulated barn to maintain a warm temperature. However, cattle are herbivores and grazing on grass

Foot and mouth disease is an infectious disease affecting cattle, sheep, pigs, goats and deer. The sale of livestock is severely restricted by an outbreak of the disease.

BSE is a neurological disease that affects cattle. On 8 March 2006 the export restrictions against export of British beef to Europe were lifted, as the feed controls and inspections had been effective in controlling the disease.

Herbicides are chemicals that are used to kill weeds so the weeds do not compete with the crop for water, light or nutrients.

Insecticides are used to kill insect pest species that eat and damage the crop plants, but insecticides can also kill beneficial insects, such as bees.

Biodiversity is the number of different types of plants and animals living and feeding in an ecosystem. These may include pollinators and biological control species that are beneficial to crops.

rather than being fed with concentrates or silage produces the best lean meat. Following scares such as BSE ('mad cow disease'), more people are concerned about intensive farming and are supporting organic and natural farming.

- Having fewer links in the food chain would increase the efficiency of food production. Humans could consume more plant proteins such as beans and eat less red meat. However, not all land used for grazing, such as hill pastures and marsh areas, is suitable for growing crops.
- Novel foods such as Quorn™ are slow to be taken up by consumers unless the foods have a niche appeal such as being low-fat or cholesterol-reducing, or are advertised by celebrities.

The costs of food production and distribution

The following points highlight some of the conflicts between the demands of the shopper for a good quality product at the lowest possible price, and the needs of the producer to increase productivity and reduce costs.

- Crops sold in supermarkets are intensively produced using fertilisers, pesticides and irrigation.
- Where greenhouses are used, there will be extra costs for the input of heat, light and carbon dioxide.
- Exotic or out-of-season crops that are grown abroad have to be flown or shipped to the UK at extra cost. People living in rural areas have the benefit of being able to purchase local produce direct from farm shops.
- Meat production costs include costs of rearing, feed material, housing, waste disposal, distribution transport to market, veterinary bills and labour.

Food distribution could also help the spread of infection. The serious outbreaks of foot and mouth disease informed the public of the distance travelled by livestock from rearing locations to fattening locations, often hundreds of miles away. It also informed them of the problems associated with live export of lambs and calves. BSE made the public aware of the risks of intensive rearing which, in the 1980s, used animal carcass protein as a protein supplement in food. This practice is now banned. Currently, we are concerned about the spread of bird flu in intensive poultry units, and poultry imports to the UK from certain countries are banned.

There are also conflicts between agricultural production methods such as the use of fertilisers, herbicides and insecticides and damage to the environment, for example excess weed growth in rivers and streams and loss of biodiversity. In addition to increasing the cost of food, distribution and transport use up non-renewable fuels.

Activity – Are shoppers becoming more aware of food production issues?

In a small group, take a large piece of paper (A3). Using a pencil and ruler measure and mark the paper into four sections, as shown below.

Write down the positive and negative effects of managing food production. For example, you might say that the use of an insecticide made sure that there were no holes in the cabbage leaves and no caterpillars lurking between the leaves!

You might like to include other topics such as selling only seasonal or local produce, genetically modified (GM) crops, organic farming, free-range chicken or pigs, set-aside, problems faced by families on very low incomes and any issues relevant to your local area. Do a web search for 'sustainable farming'; the following website might be useful: www.sustainweb.org.

Food production	
Positive aspects	Negative aspects
Food distribution	
Positive aspects	Negative aspects

What happens to the waste material produced by plants and animals?

Ask yourself what happens to all the dead leaves, dead plants and animals, piles of faeces and other natural debris? What causes them to decay?

In Chapter 3 you saw that in a rainforest there was rapid recycling of nutrients. If the trees, which were the source of the leaf litter, were removed the cycle stopped. Here we will look in detail at the nutrient cycling process.

Dead leaves, other dead organic matter and animal waste products are broken down or decomposed by bacteria and fungi (**microorganisms** called **decomposers**). If you turn over a damp leaf on the ground in autumn you might see fine cotton-wool like threads on the underside. These are the fungal hyphae (feeding structures of the fungus), which are chemically breaking down, or digesting, the dead leaf cells (enzymes and digestion are described in Sections 7.4 and 7.5). The cellulose of plant cell walls is very difficult to break down and so the enzyme cellulase secreted by the fungi is very important in starting the decay process.

> Bacteria and fungi are examples of **microorganisms**.
>
> Microorganisms involved in the process of decay are called **decomposers**.

10. Dead leaves have cellulose cell walls. Name the types of enzymes that will be required to break the leaves down.
11. Name the nutrient that is taken up and combined with sugars to make amino acids, which can then be assembled into proteins.
12. Name the mineral ion which is taken up from soil and used to make chlorophyll.

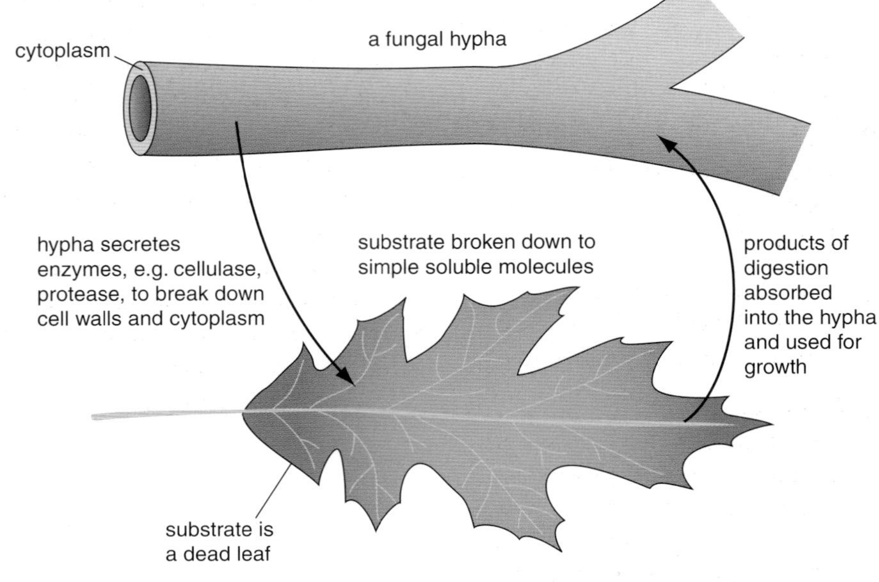

Figure 6.21 Fungi carry out extracellular digestion to break down dead leaves.

Water – **O**xygen – **W**armth

Microorganisms require some of the same conditions that you do – water, oxygen, warmth and nutrients.

The digestion reactions take place faster if there is warmth and oxygen. These are the conditions you will find in a garden compost bin or a municipal waste composter.

The breakdown of waste materials by microorganisms produces simpler, soluble materials that can be used as nutrients by plants. In a stable ecosystem, some nutrients released by the decay process are used for bacteria or fungal growth and some are taken up by the trees to be used again to make more leaves – the nutrients are cycled.

6.8 How do carbon atoms move around in the carbon cycle?

The **carbon cycle** describes how carbon atoms circulate through carbon compounds in living organisms to carbon dioxide in the air and back again.

Carbon dioxide makes up 0.037% of the atmosphere by volume and is the raw material for photosynthesis in plants (see Section 6.1). Let's examine the possible route taken by a carbon atom in a carbon dioxide molecule.

- Carbon dioxide is combined with hydrogen from water, using energy from sunlight, in the process of photosynthesis.
- The glucose formed by photosynthesis can be used to synthesise structural and storage materials in the plant (see Figure 6.8). In this way the carbon atoms from carbon dioxide become incorporated into oils, starch, cellulose and proteins in the plant biomass.
- If the plant is eaten by a herbivore, the carbon atoms can become incorporated into the fat, glycogen or proteins in the animal, and may pass to the next animal in the food chain, a carnivore. This chain – plant to herbivore to carnivore – can be thought of as the 'light chain' because it starts with the capture of light (energy).

13. What is the only process in the carbon cycle that removes carbon dioxide from the atmosphere?
14. Which three groups of living organisms are shown respiring in the carbon cycle diagram?
15. What additional source of carbon dioxide is increasing atmospheric carbon dioxide?
16. Complete the cycle for a carbon atom as it passes from the atmosphere to its eventual release back into the atmosphere by respiration in a rabbit. Name the processes and structures in which it will be involved.
    - Diffuses as carbon dioxide through open stomata into leaf.
    - _______ in leaf.
    - Stored as _______ and used to make a structural material, _______.
    - Eaten by rabbit and _______ in rabbit's gut by enzymes.
    - Taken up as _______ into the rabbit's bloodstream.
    - _______ in rabbit muscle cells.
17. a) Name the two types of decomposer organisms in the carbon cycle.
    b) Decomposers break down dead organic matter into small soluble molecules. Give two different uses for these molecules.

- Dead plant or animal remains and waste material pass to decomposers, bacteria and fungi in the soil or leaf litter. This part of the cycle can be considered as the 'dark chain' and is very important in terms of cycling nutrients and energy. The bacteria and fungi secrete enzymes onto the dead organic matter to break it down (digest it). Some of the soluble products of digestion are taken up by the bacteria and fungi for their growth or respiration. The remainder stay in the soil and can be taken up by plants. The nutrients have been cycled.
- Respiration of plants, animals and decomposers returns the carbon atoms to the atmosphere as carbon dioxide.
- There can be an extra input of carbon dioxide into the atmosphere by the burning of fossil fuels such as diesel, petrol, coal, oil and natural gas and by burning wood.

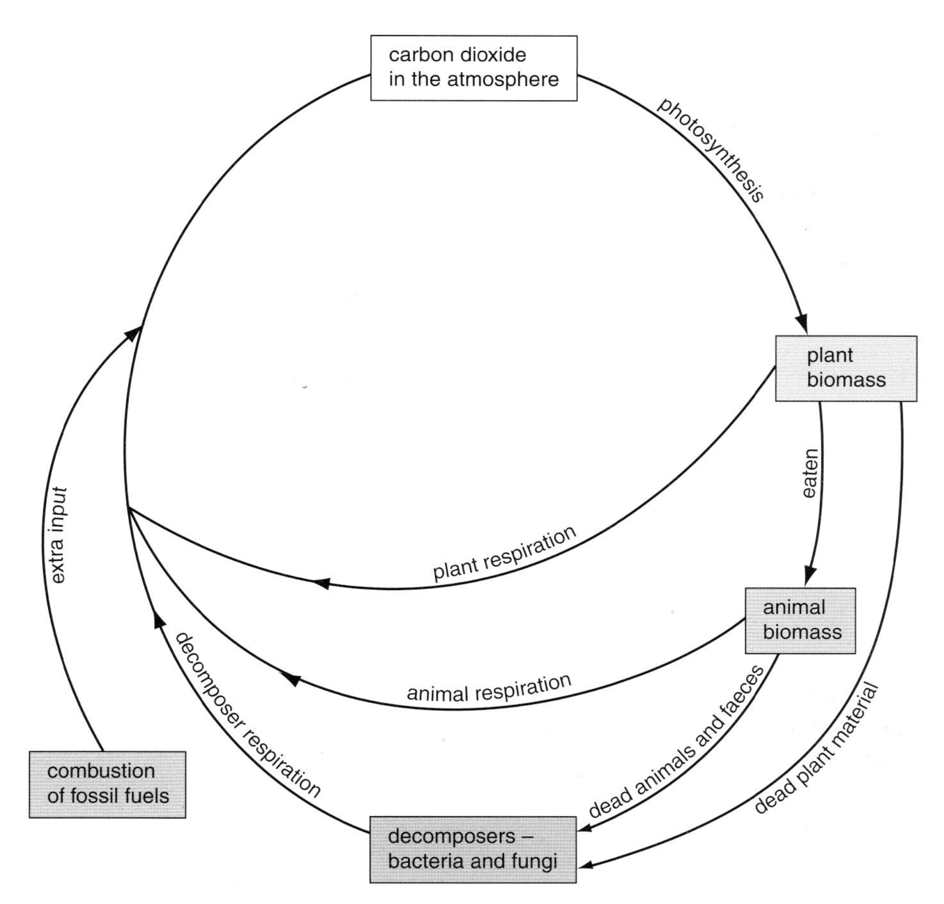

Figure 6.22 The carbon cycle

We have looked at the cycling of carbon atoms and at how nutrients are released and cycled by microorganisms. In the **carbon cycle** (Figure 6.22) the arrows go around in a circle, showing the route of the carbon. However, energy flows through an ecosystem in a straight line. Energy enters the ecosystem, is converted and used, and some passes out of the ecosystem – it cannot be cycled.

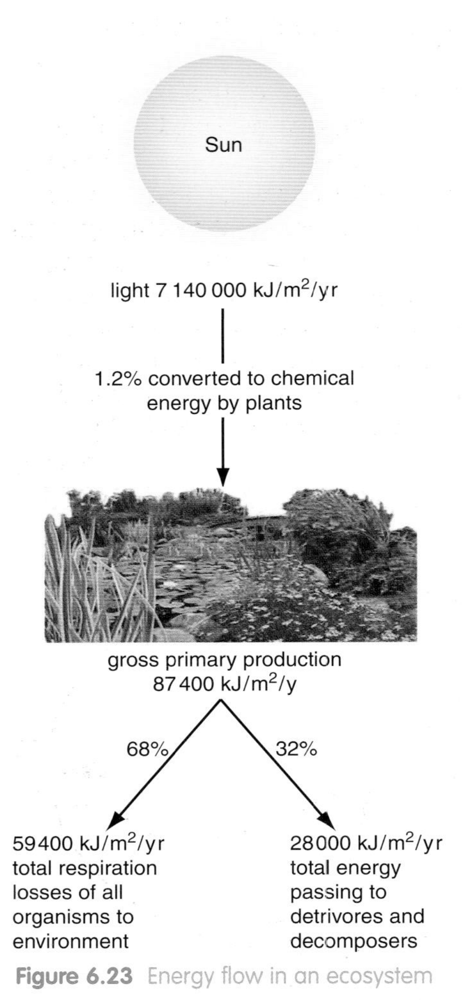

Figure 6.23 Energy flow in an ecosystem

In the process of photosynthesis, the energy from sunlight is used by plants to convert carbon dioxide and water into chemical potential energy in sugar. In the plant, or the animal that eats the plant, respiration releases the chemical energy in the sugar molecules and heat is lost from the organism. This energy flow is summarised in Figure 6.23.

This can also be written as:

light → chemical energy in food → heat lost to the environment

Summary

- ✓ **Photosynthesis** is the process in green plants that converts light energy into chemical energy.
- ✓ **Chlorophyll** in chloroplasts absorbs light (energy).
- ✓ The raw materials for photosynthesis are carbon dioxide and water. The products are oxygen and glucose.
- ✓ The glucose can be used for energy, combined with nitrates to make plant proteins, converted to starch for storage, or used as a structural material such as cellulose.
- ✓ If carbon dioxide is in short supply or the temperature or **light intensity** is low, the rate of photosynthesis will be reduced. Any of these factors could be a **limiting factor**.
- ✓ Plant roots absorb mineral ions, including nitrates (needed to form amino acids) and magnesium (needed to form chlorophyll). Mineral deficiency can cause stunted growth and yellow leaves.
- ✓ To maximise greenhouse crop production, light intensity, carbon dioxide concentration and temperature are artificially controlled at optimum levels.
- ✓ A **pyramid of biomass** shows the mass of organisms at each **trophic** (feeding) **level**. It measures the material present at the time.
- ✓ In a pyramid of biomass there are generally large numbers of small organisms at the base, but size of organisms increases and numbers of organisms decrease at each trophic level.
- ✓ In a **pyramid of energy** there is less energy in each successive level.
- ✓ Energy transferred from one level to the next is reduced because some energy is used for movement or is lost in waste materials, and in mammals and birds heat is lost in maintaining body temperature above that of the environment.
- ✓ Food production would be more energy efficient if humans fed as herbivores, because this reduces the number of stages in the food chain.
- ✓ The efficiency of food production is increased by restricting movement and heat loss in food animals.
- ✓ Bacteria and fungi play an essential role in decomposing dead organic matter and in recycling mineral ions to the plants.
- ✓ Bacteria and fungi (**decomposers**) break down dead organic matter by extracellular digestion.
- ✓ Nutrients cycle within and are reused in an ecosystem – there is a finite supply of nutrients.
- ✓ The carbon cycle shows how carbon atoms circulate through carbon compounds in living organisms to carbon dioxide in the air and back again.
- ✓ Energy flows through an ecosystem – the light (energy) input is available every day.

1. A glasshouse owner wishes to grow many successive crops of lettuce and harvest the crops all year round.
   a) Examine the computer control panel in Figure 6.24 and decide which factors are limiting the productivity. *(3 marks)*
   b) Suggest suitable new settings. *(3 marks)*
   c) The grower decides to install a paraffin heater. What problems might this cause? *(2 marks)*

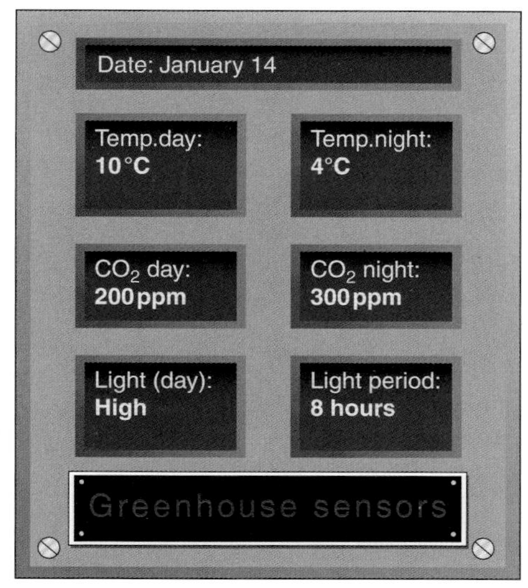

Figure 6.24

2. A student is planning a trial to find the best concentration at which to apply a liquid fertiliser. The recommended strength is 20 cm^3 per 10 dm^3 – which the student calls 100% strength. He decides to try the first concentration at double the recommended strength, i.e. 200%, and then a range of other concentrations. He records the dilutions in Table 6.3. (1 dm^3 = 1000 cm^3)

Volume concentrated solution in cm^3	Volume water in cm^3	Final concentration in %
40	9 960	200
20	9 980	100
10	9 990	
5	9 995	
2.5	9 997.5	
0	10 000	

Table 6.3

   a) Complete the table to find his dilutions. *(4 marks)*
   b) Do you think the range of percentage concentrations chosen was suitable? Give a reason for your answer. *(2 marks)*
   c) The student's idea was to harvest the plants at the end of the experiment by cutting them off at soil level to find the fresh biomass. His friend suggested that he should harvest the plants and then find the dry biomass. What is the difference between fresh and dry biomass? Which do you think would give the most accurate result, and why? *(3 marks)*

3. Look at the carbon cycle in Figure 6.25.

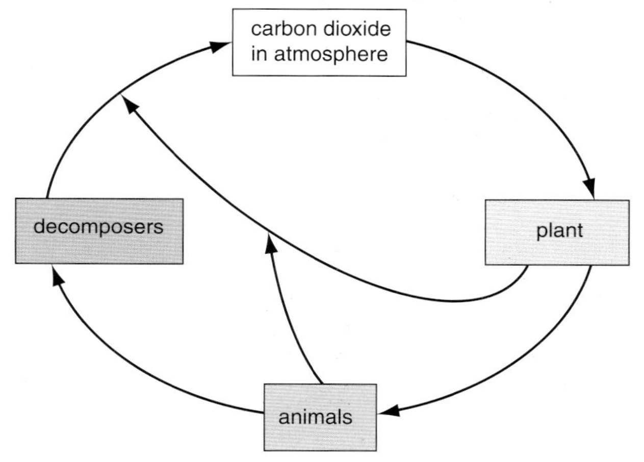

Figure 6.25 Carbon cycle

   a) i) Name the only type of organism shown which removes carbon dioxide from the atmosphere. *(1 mark)*
      ii) For what process is the carbon dioxide used? *(1 mark)*
   b) Name i) a structural material and ii) a storage product in a plant, which contain carbon. *(2 marks)*
   c) If the cow eats grass it assimilates the carbon atoms into compounds in its body. Name i) a structural material and ii) a storage product in a cow, which contain carbon. *(2 marks)*
   d) The manure (cow pat) of the cow contains the remains of the grass. What is the general name given to organisms that help in the chemical breakdown of the manure? *(1 mark)*

4 A student turns over a dead log and finds fine cotton-like threads on the underside. Complete the sentences below to explain what is happening at a), b) and c) in Figure 6.26.

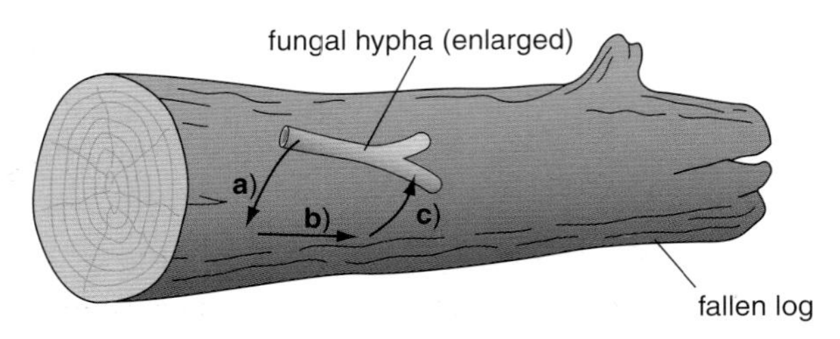

Figure 6.26

a) Enzymes are secreted to ______ the wood. These enzymes will include ______.
b) Soluble ______ and ______ produced by the enzymes will be taken up by plants.
c) The soluble products will be used by the plant for ______. *(5 marks)*

5 The mangrove swamps in the Gulf of Mexico support five trophic levels.

The main producer in this ecosystem is the seagrass, which grows underwater and is consumed by small herbivorous fish.

Unicellular algae make up a smaller biomass but they are the main food of the zooplankton (microscopic animals).

The zooplankton provide the food source for the invertebrates such as shrimp and for small carnivorous fish.

Puffer fish eat mainly the invertebrates. Together with the larger fish they make up the fourth trophic level.

The top carnivore is the snapper fish.

Table 6.4 shows the biomass at each trophic level.

	Biomass in kg/km^2
Primary producers	3700
Primary consumers	39
Secondary consumers	18
Third consumers	6
Top carnivore	1

Table 6.4

a) Use the data in Table 6.4 to construct a pyramid of biomass for this ecosystem on graph paper. Use a square 1 cm × 1 cm (25 small squares on 2 mm graph paper) to represent a biomass of 50 kg/km^2; one small square 2 mm × 2 mm therefore represents a biomass of 2 kg/km^2.
b) Label the organisms at each trophic level beside your diagram. *(6 marks)*
c) What do you notice about the area representing the primary producers? *(1 mark)*
d) What can you say about the change in biomass at each trophic level? *(1 mark)*
e) The biomass given is what is growing at a given time. The algae reproduce every few days. How would this contribute to the food available to the zooplankton? *(1 mark)*
f) The list of species takes no account of bird predators flying in to catch and eat the fish. Suggest which trophic level might be most reduced by these predators. *(1 mark)*

Chapter 7
Enzymes: how do they function and how can their properties be used industrially?

At the end of this chapter you should:

- ✓ know that enzymes are proteins which act as biological catalysts that increase the rate of reactions in living organisms;
- ✓ understand that protein molecules and enzymes are made of chains of amino acids coiled together;
- ✓ be able to explain why the shape of an enzyme is vital for its function;
- ✓ be able to explain why heat affects enzyme function;
- ✓ know that enzymes have an optimum temperature and optimum pH;
- ✓ know that enzymes catalyse the reactions in processes such as respiration, photosynthesis and protein synthesis;
- ✓ be able to summarise aerobic respiration and explain how energy is released by a chemical reaction inside mitochondria;
- ✓ explain how amino acids and proteins are synthesised in reactions catalysed by enzymes;
- ✓ be able to describe the digestive enzymes that catalyse the breakdown of different foods, and where this happens;
- ✓ be able to describe how enzymes produced by microorganisms are used in detergents to remove food stains;
- ✓ be able to describe applications of enzymes used in industry including pre-digested baby food and products containing sugar or fructose syrup;
- ✓ be able to evaluate the advantages and disadvantages of using enzymes in the home and in industry.

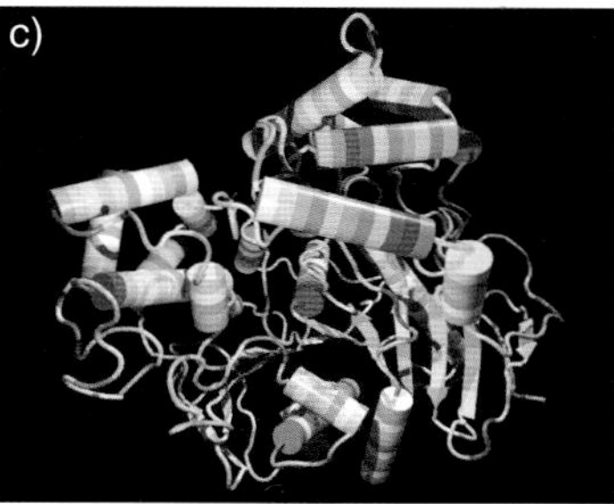

Figure 7.1 a) Why is biological washing powder called biological when it's full of chemicals? b) Biological detergents work really well on food stains by literally digesting the food with enzymes just like the enzymes in your digestive system. c) A molecule of the fat-digesting enzyme lipase.

7.1 What are enzymes and what are their functions?

Enzymes are biological molecules that catalyse reactions in living organisms.

A **substrate** is the substance acted on by an enzyme.

❶ What is the benefit of enzyme-controlled reactions inside living cells?

❷ Give three advantages of using enzymes in biotechnology applications.

❸ Give two similarities and two differences between enzymes and other catalysts.

Enzymes are called biological catalysts because they control most of the reactions that take place in living organisms. All catalysts change the rate of chemical reactions but are not themselves changed by the reaction. Enzymes catalyse essential reactions such as respiration and photosynthesis that take place inside cells. The enzyme is not used up so it can catalyse the same reaction many times.

The function of enzymes is to enable reactions to take place at lower temperatures. At low temperatures, such as the temperature of the human body, the chemical reactions inside cells would take place very slowly without enzymes.

Enzymes are used in many processes in the food and biotechnology industries so that the reaction temperature can be kept low. This saves energy used to heat the container and so saves money (see Section 7.5). Enzymes also increase the reaction rate so that large amounts of product can be made quickly.

Enzymes differ from chemical catalysts because they are protein molecules. Therefore, they have some properties of both catalysts and proteins. As catalysts, enzymes change the rate of reactions and as proteins they have a complex 3-D structure that allows other molecules to fit into the enzyme.

7.2 How do enzymes work?

Cracking the name code for enzymes. The ending *-ase* tells you that the substance is an enzyme. The first part of the name tells you the substrate. For example, protease is the enzyme that breaks down protein, and amylase breaks down amylose. Think of some compounds with names ending in *-ose*. What type of substance are these?

Each enzyme works on only one particular reaction, to produce certain products from one **substrate**. You probably know that when your body digests food, different enzymes work on different food types. Protease enzymes speed up digestion of protein and amylase breaks down the substrate amylose (commonly called starch) to give the products glucose and fructose.

Why does an enzyme only work on one substrate? Let us look at the structure of enzymes in more detail. You will need to think about the three-dimensional structure of the molecule. Remember that enzymes are protein molecules.

Amino acids are the building blocks of proteins. They are joined end to end to form a long chain. This molecular chain is stabilised when it is folded into a three-dimensional shape and cross-bonds form between the amino acids (rather as a girl might twist up her hair and put in slides). All enzymes have this same basic structure, but they have a different sequence of amino acids to form different protein molecules. The order

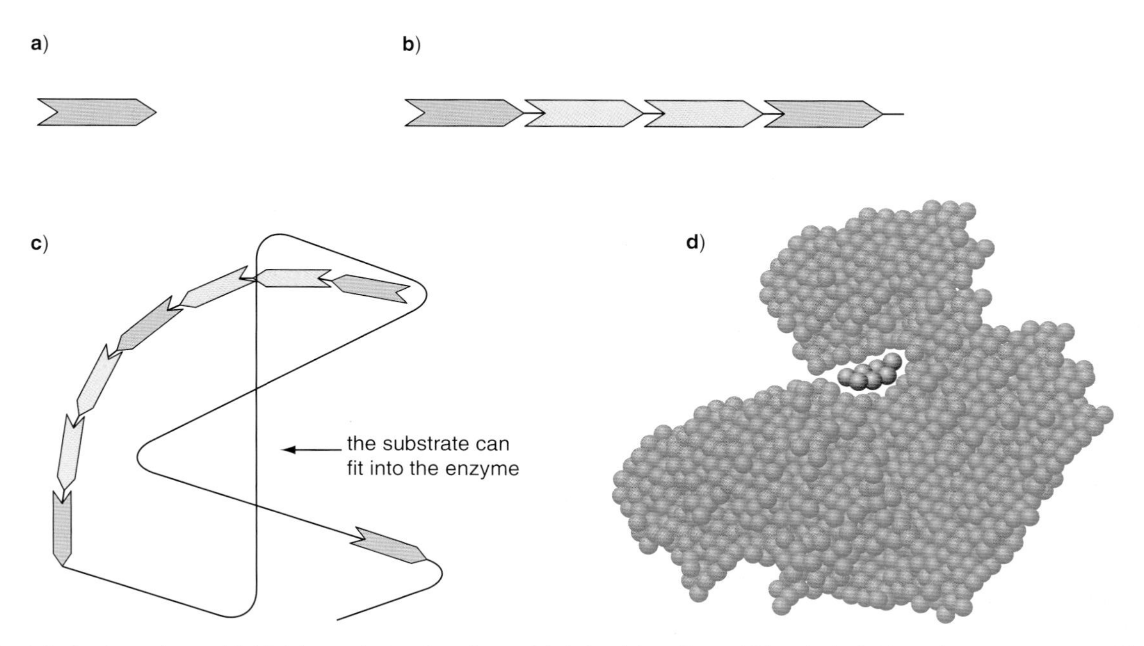

Figure 7.2 a) A single amino acid. b) A long chain of amino acids joined together. c) The chain folds to become more stable. d) The complete three-dimensional enzyme with substrate.

A **model** is a simplified picture that scientists construct to fit observed data to a theory.

The **lock and key model** suggests that each enzyme (the lock) has a special shape, which means that only one substrate (the key) can fit into each enzyme.

The **activation energy** is the energy that must be provided to the reactants to start a chemical reaction.

in which amino acids are joined together is controlled by the DNA code in the nucleus. Each tightly folded protein molecule or enzyme has ridges and grooves, giving it a unique shape. This means that each enzyme can fit only one type of substrate.

This idea of the substrate fitting into the enzyme is called the **'lock and key' model** for enzyme structure. Enzymes unlock a particular reactant molecule so that a chemical reaction can take place and the molecular bonds can be broken. This is why each enzyme is specific to one type of chemical reaction.

What conditions affect enzymes?

Rates of enzyme reactions increase with increase in temperature. To understand why, you can apply your understanding of what is happening at the molecular level. First, the enzyme and substrate molecules must collide with enough energy for the substrate to fit into the enzyme structure. This is called the **activation energy**. This is a bit like putting a missing piece into a jigsaw puzzle: you can place a missing piece over a gap, but it needs a little push (energy) to fit it in.

An increase in temperature will cause the particles to move faster, that is to increase their kinetic energy. As a result enzyme molecules and substrate molecules will collide more often and with more energy. The enzyme can catalyse more reactions so the reaction rate is increased.

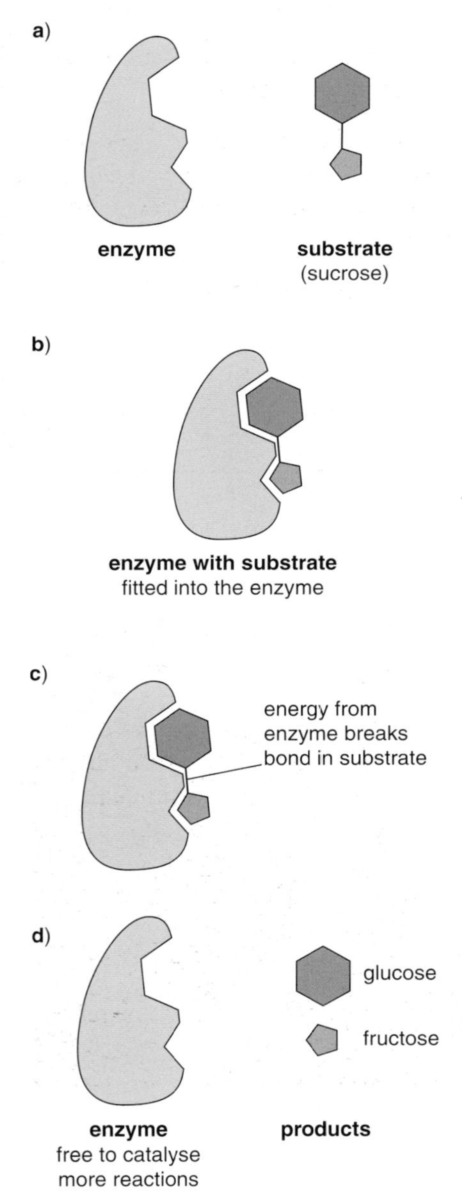

Figure 7.3 This sequence of diagrams shows how an enzyme breaks down a substrate. Here, the substrate is sucrose and the products are glucose and fructose.

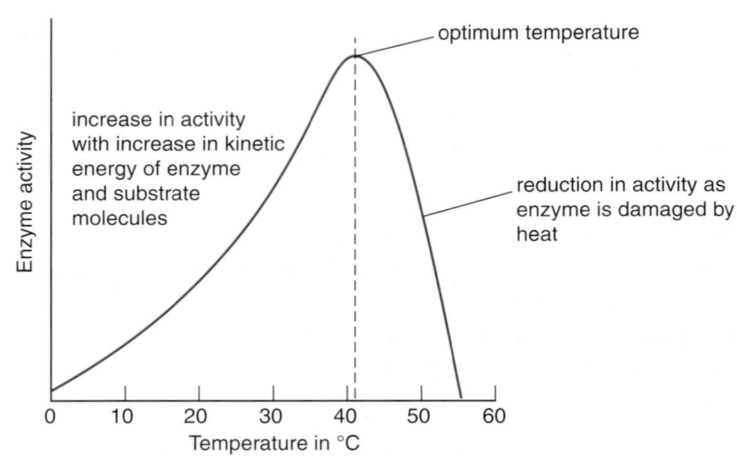

Figure 7.4 Sketch graph to show the effect of heat on an enzyme reaction

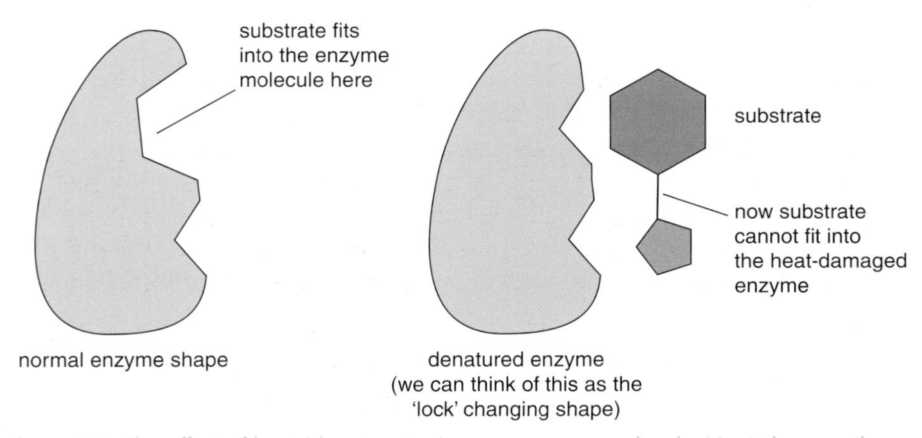

Figure 7.5 The effect of heat (denaturation) on an enzyme molecule. Heat changes the bonds and changes the three-dimensional shape.

But this is where enzymes differ from catalysts: enzymes function best at an optimum temperature (Figure 7.4). An increase in temperature above this changes the shape of the protein molecule (Figure 7.5). You can see the effect of change in the protein structure when you cook an egg white. When the enzyme molecule shape has been changed, the substrate can no longer fit into the enzyme – we say that the enzyme has been **denatured** and it can no longer function as a catalyst.

Living organisms are adapted to exist in a range of habitats, so the enzymes inside bacteria which live in hot springs will be able to function within a different temperature range from those enzymes in plants in the Arctic. There are also differences in pH levels at which enzymes work. Each enzyme has an optimum pH, which is the pH at which it functions best or most effectively. These differences have been used in industrial applications such as those discussed in Section 7.5.

When temperature destroys the three-dimensional shape of a protein molecule or enzyme, it is **denatured**.

4 Why are enzyme reactions slow at low temperatures?

5 Why does the rate of product formation increase as the temperature increases between 5 °C and 30 °C in enzyme reactions?

6 A student was carrying out an experiment using the enzyme protease on egg white. He found that the white solution became colourless for all temperatures up to 45 °C, but above this temperature the solution remained cloudy. Can you explain this?

7 What term is given to the temperature at which enzymes function most efficiently?

8 The mouth is slightly alkaline but the stomach has a low pH. Why do you think salivary amylase stops working when the food it is mixed with enters the stomach?

9 Why can protease break down protein but not starch?

Activity – Who was Emil Fischer?

Emil Fischer was born in Germany in 1852. His father, who said that Emil was too stupid to be a businessman and had better be a student, sent Emil to the University of Bonn in 1871 to study chemistry. Emil Fischer won a Nobel prize in 1902. This was the first prize awarded for organic chemistry. He was famous for rigorous teaching and making links between science and industry. Six of his students later went on to become Nobel prize winners.

Figure 7.6 Emil Fischer

Emil Fischer first suggested the 'lock and key' model in 1894. The amazing thing about Fischer's model was that he suggested this before scientists had discovered that enzymes were protein molecules.

The lock and key model is used today to develop computer simulations that investigate possible new drugs. The computer programs design and test thousands of different molecules with different shapes, to see how they could 'fix' onto receptors on the surface of target disease cells.

1 Write and illustrate a newspaper report about Fischer's great award. Find out about him by searching for his name and biography. How long after his discoveries was he rewarded?

2 Find the answers to the following questions using books or an internet search.
   a) In Emil Fischer's model, what does the lock represent?
   b) After the 'lock and key' model was suggested, how long was it before the nature of enzymes was worked out, and how long before a three-dimensional model of a protein was first made?

3 Draw a diagram to show the 'lock and key model' as a lock and key, and a more modern diagram showing an enzyme and substrate.

What reactions do enzymes control inside cells?

Most chemical reactions in the cell take place in the cytoplasm. Different enzymes catalyse each of the chemical reactions in the cell. Enzymes working inside cells have a constant temperature and pH.

Building protein for growth

Protein synthesis is the building of protein molecules from a chain of amino acid molecules, using enzymes to catalyse the reaction.

Aerobic respiration occurs when oxygen reacts with glucose and releases energy. Carbon dioxide and water are the products of the reaction.

Growing cells need to build new structures inside the cell and develop more organelles such as mitochondria. To do this they make proteins. Amino acids can be joined together to form structural proteins and more enzymes. Enzymes catalyse the reactions that make proteins inside the cell. This is called **protein synthesis** – different enzymes control which proteins can be built. Different cells do different jobs, and so there are thousands of different proteins.

Respiration to release energy

Making larger molecules from smaller molecules needs energy. All cells in living organisms release energy from food by respiring. **Aerobic respiration** is a reaction in cells that uses glucose and oxygen to release energy. Aerobic respiration occurs as a series of reactions, each controlled by an enzyme.

The overall reactions in aerobic respiration are summarised by the equation:

glucose + oxygen → carbon dioxide + water (+ energy)

$$C_6H_{12}O_6 + 6O_2 \rightarrow 6CO_2 + 6H_2O \text{ (+ energy)}$$

First, glucose is broken down in the cytoplasm to form smaller molecules that can enter the mitochondria. In the mitochondria a sequence of reactions takes place. In the final stage hydrogen atoms removed from the glucose are combined with oxygen to form water and energy is released.

So why is respiration a series of reactions and not a single reaction? If you oxidise glucose by setting fire to it you end up with the same overall reaction and the same products, but a great deal of energy is released rapidly producing heat – we call this combustion. This type of reaction within cells would damage the proteins. Instead, in respiration the energy is released bit by bit at each reaction, in small units that have sufficient energy to power the other reactions in the cell. You could compare this to arriving at a holiday destination with all your money as large value notes – it's best if you have smaller values and coins for buying drinks and chocolates.

10. Name two large molecules made in plant cells by joining together smaller molecules.
11. In which part of the cell is most of the energy in respiration released?
12. Combustion is one reaction but respiration takes place as a series of small reactions. What points are the same about these reactions and how do they differ?

What do cells do with these small units of energy?

Remember that energy is needed for building up small molecules into large molecules, such as amino acids into large protein molecules. Other chemical reactions in the cytoplasm also need energy. For example, plants need energy to add nitrates to the sugars produced by photosynthesis and form amino acids. Plants also build up glucose molecules into cellulose for their cell walls.

Photosynthesis itself is a chemical reaction, and without plant enzymes this reaction could not take place. In photosynthesis the enzyme systems in chlorophyll convert light (energy) into chemical energy in the form of glucose. The majority of food webs are based upon this reaction (see Chapter 6).

Energy released by respiration is used by warm-blooded animals to maintain their body temperature, and for muscle fibre contractions so that the animal can move.

13. Examine the following statements and identify which are correct.
    a) Plants respire at night and photosynthesise in daylight.
    b) Plants respire for 24 hours a day.
    c) Plants photosynthesise for 24 hours a day.
    d) For plants to grow, the rate of photosynthesis must be greater than the rate of respiration over the year.
    e) For growing plants the rate of photosynthesis equals the rate of respiration over the year.
    f) Plants can photosynthesise for longer each day in spring than in winter.
14. Suggest three ways in which animals make use of the energy released in respiration.

Are enzymes used outside cells in the body?

Digestive enzymes catalyse the breakdown of food as it passes down the intestine from the mouth. The digestive system can be thought of as a tube through the body carrying food.

A **gland** is an organ that produces and releases (secretes) a substance used by other cells.

Digestive enzymes are produced by specialised cells in **glands**, such as the salivary gland or the pancreas, or by cells in the lining of the stomach and intestine. These enzymes pass out of the cells and are released into the intestine.

The purpose of digestion is to break down large insoluble molecules into small soluble molecules. For example, amylase breaks down the substrate amylose (starch) to give glucose and fructose. Smaller

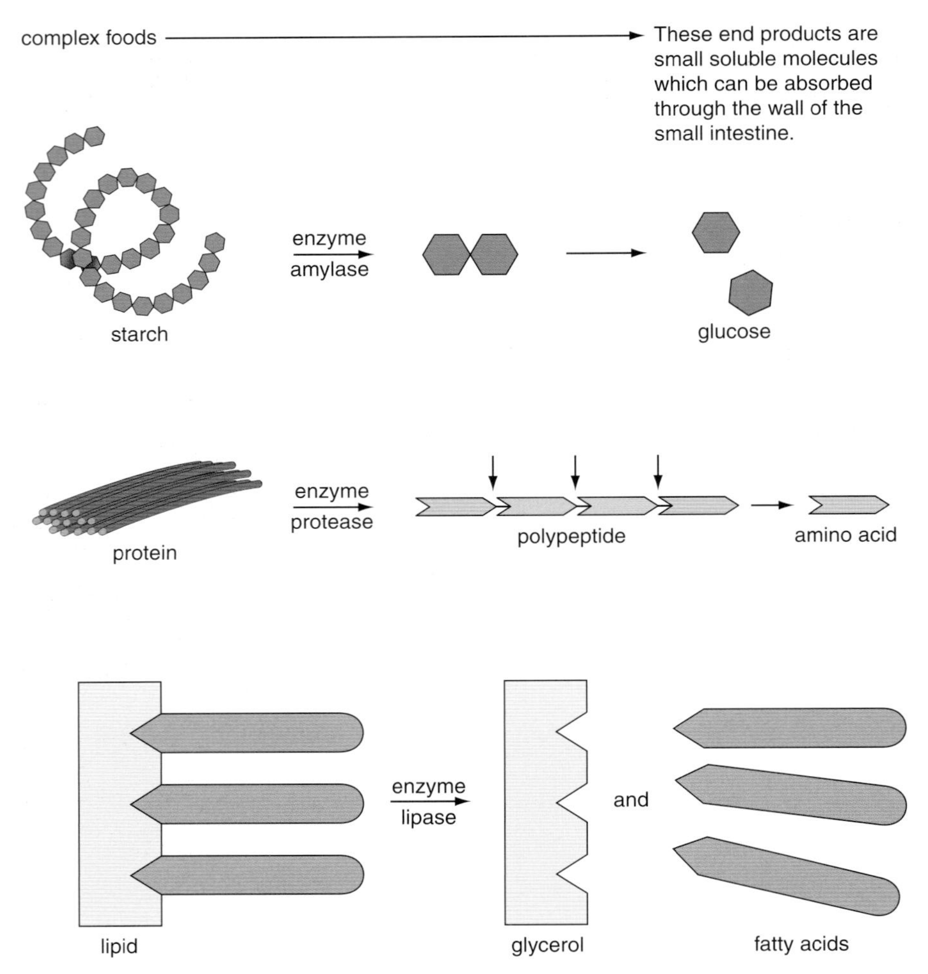

Figure 7.7 The breakdown of the complex food molecules into small, soluble, usable molecules

molecules are needed because they are soluble and can pass (diffuse) from the small intestine into the blood plasma for transport to body cells. Smaller molecules, like glucose and amino acids, are also in a form that cells can use.

Protease enzymes that break down proteins are produced by cells in the stomach lining, so protein digestion begins in the stomach. However, different proteases are produced by the pancreas and secreted into the small intestine where proteins continue to be broken down into amino acids. The stomach protease and the pancreatic protease function at different pH levels as you can see in Table 7.1.

Bile is a substance produced by the liver and released during digestion into the small intestine to provide alkaline conditions. Bile also reduces the size of fatty droplets, making them easier to digest.

Other substances are released that help the enzymes to function by providing the optimum pH. Hydrochloric acid is released into the stomach to provide the low pH required by the protein-digesting enzyme called protease. **Bile** produced in the liver cells is stored in the gall bladder and added to the small intestine via a small duct (tube). Bile neutralises the acidic food, leaving the stomach to provide the higher pH needed by enzymes which work in the small intestine.

Amylase is the digestive enzyme that breaks down starch into glucose and fructose.

Lipase is the digestive enzyme that breaks down lipids (fats and oils) into fatty acids and glycerol.

Protease is the digestive enzyme that breaks down proteins into amino acids.

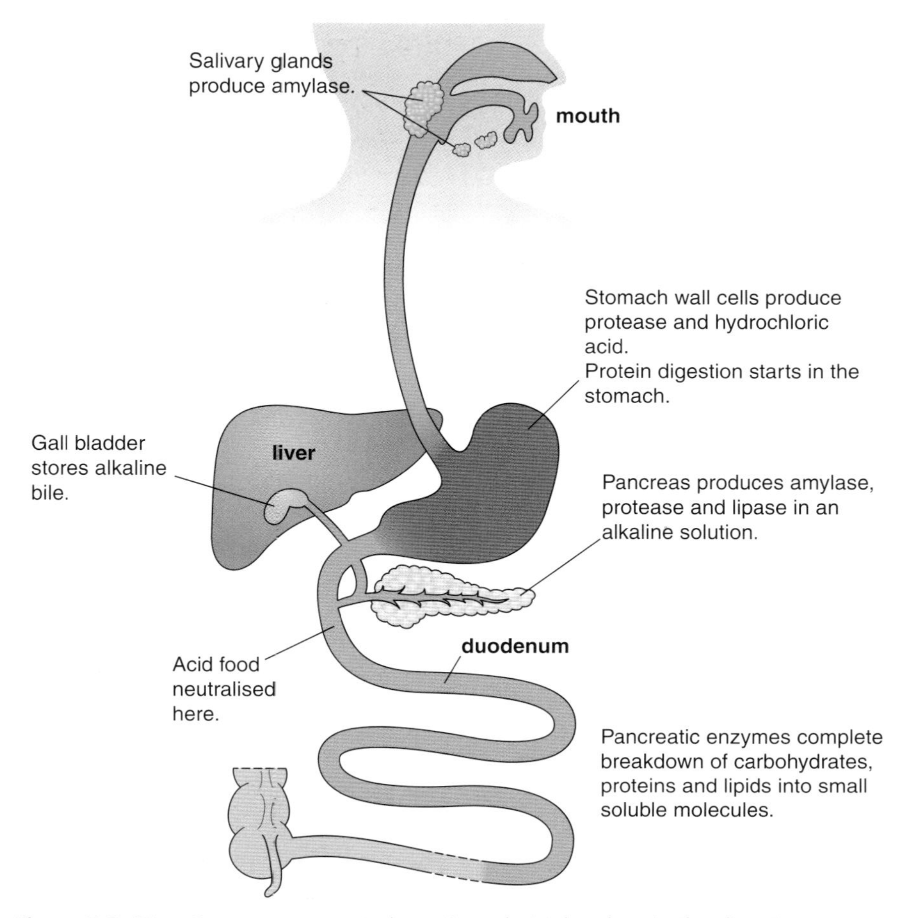

Figure 7.8 Digestive enzymes control reactions that take place in the digestive system.

Enzyme	Site of production	Site of action	pH	Substrate	Products
Salivary amylase	Salivary glands	Mouth	8	Starch	Sugars
Pancreatic amylase	Pancreas	Small intestine	8	Starch	Sugars
Protease	Cells in stomach wall	Stomach	2 (hydrochloric acid produced by the stomach acidifies food)	Protein	Polypeptides and amino acids
Protease	Pancreas	Small intestine	8 (bile from the liver neutralises the acid food leaving the stomach)	Protein and polypeptides	Polypeptides and amino acids
Protease	Small intestine wall	Small intestine	8	Peptides	Amino acids
Lipase	Pancreas and cells in small intestine wall	Small intestine	8	Lipids	Fatty acids and glycerol

Table 7.1 Enzymes of the digestive system

15 a) The food leaving the mouth is slightly alkaline. How is the pH lowered in the stomach?
b) What is the optimum pH for the enzyme amylase?
c) The food leaving the stomach is strongly acidic. How is the pH raised?
d) What is the optimum pH for all the pancreatic enzymes?
e) What is the optimum pH for stomach protease?

16 At the start of the chapter, in question 1, 'What is the benefit of enzyme-controlled reactions inside living cells?', you considered temperature. Now think about the rate of energy production and give a second answer. Suggest how very slow uncatalysed reactions would affect life.

17 Copy and complete the four blanks in the table.

Enzyme	Large complex molecules	Small soluble end products of digestion
Amylase	Starch	a) _______
Protease	b) _______	Amino acids
c) _______	Lipids (fats and oils)	d) _______ and glycerol

Table 7.2

18 a) Name the two places where amylase is produced.
b) Name the two sites of protein digestion.
c) Describe where fats and oils are digested.

19 a) Where is bile produced?
b) Where is bile stored?
c) What is the function of bile?

Industrial production and use of enzymes

The list of industrial applications for enzymes is amazing and includes food, wine and beer production, making fructose sweeteners for soft drinks, laundry detergents, producing the 'stone-washed' jeans effect and manufacturing pharmaceuticals. The food industry uses many enzymes, which can be 'tailor-made' to give an exact product for the development of a new food.

Many industrial enzymes come from soil microorganisms. One of the best sources of industrial proteases used in laundry detergents is *Bacillus* species, a common soil bacterium.

Like all living organisms, bacteria and fungi produce and secrete enzymes. The enzymes pass out of their cells into the environment. The enzyme products, small soluble molecules, are then taken into the organism's cytoplasm (see Section 6.7, Figure 6.21). Microorganisms are collected from different locations around the world and the enzymes they produce are tested. Biotechnologists look for enzymes that could be used for particular processes, and that work at the temperature of the industrial plant.

As bacteria and fungi normally live in cool environments, their enzymes usually function at low temperatures. However, if an enzyme is required in industry to work at high temperature or extremes of pH, then enzymes have to be found from microorganisms living in those environmental conditions. Enzymes that act on the right substrates but at the wrong temperature may need to be modified by gene transfer.

A **fermenter** is a large steel vessel used for biochemical reactions. Sensors monitor the conditions inside. The sensors send information to a computer, which then controls input valves to maintain the temperature, pH, nutrient and oxygen levels at the optimum value.

To produce the enzymes, the microorganisms are cultured in a fermentation process. At the end of the batch process, the **fermenter** is emptied and the enzymes are collected from the solution in which the bacteria have grown.

Figure 7.9 An industrial fermenter for producing enzymes. The raw material is converted by the enzymes as it flows slowly over beads with enzymes trapped on the surface (see Figure 7.10).

Carbohydrase enzymes break down complex carbohydrate molecules such as starch into simple sugars.

Isomerase is an enzyme that converts glucose into fructose.

Enzymes in the food industry

An early application of enzymes in the food industry was the production of sweet syrups by breaking down starch. This process can be done chemically by boiling starch with acid, but the reactions using the enzyme method give a pure and reliable product, with no other by-products. Another advantage is that energy costs are lower as the enzyme process can be carried out at a lower temperature.

Glucose syrup is widely used as an ingredient and food additive. Most glucose syrup is produced from maize (corn) using enzymes. Maize processing results in a large amount of starch waste products. Using a common and cheap (waste) material as the raw material for glucose syrup is cost effective.

Many **carbohydrase** enzymes are used in sequence to produce a variety of syrups of different sweetness. For example:

- maize starch is treated with one form of amylase to convert it to a thick starch paste;
- this paste is then reacted with different amylase enzymes to form sugars such as maltose or glucose. The reaction can be stopped here; or
- the final stage is the conversion of glucose syrup to fructose syrup, using the enzyme **isomerase**.

In the early 1970s a continuous flow system was developed in which glucose solution was added through the top and fructose solution was delivered from the bottom. The continuous process can run for about 6 months non-stop. As enzymes are proteins, their structure eventually breaks down and the enzyme needs replacing. The old enzymes are biodegradable in the environment so do not result in toxic waste. Fructose syrup is much sweeter than sucrose or glucose syrup so it can be used in smaller quantities as a sweetener. This saves food manufacturers money, but another result was that in the 1970s fructose syrup began to be used in slimming foods to give sweetness with fewer calories.

Although there are clear economic advantages of using enzymes in industrial processes, there are some disadvantages:

- development and production of new enzymes is costly;
- enzyme action is highly specific, so they have only one use;
- enzymes require exact conditions of temperature and pH in which to work;
- as enzymes are proteins, they can cause allergic reactions, and so must be encapsulated to minimise the risk of skin contact or inhalation;
- some people may object to genetically modified (GM) enzymes being used to produce food.

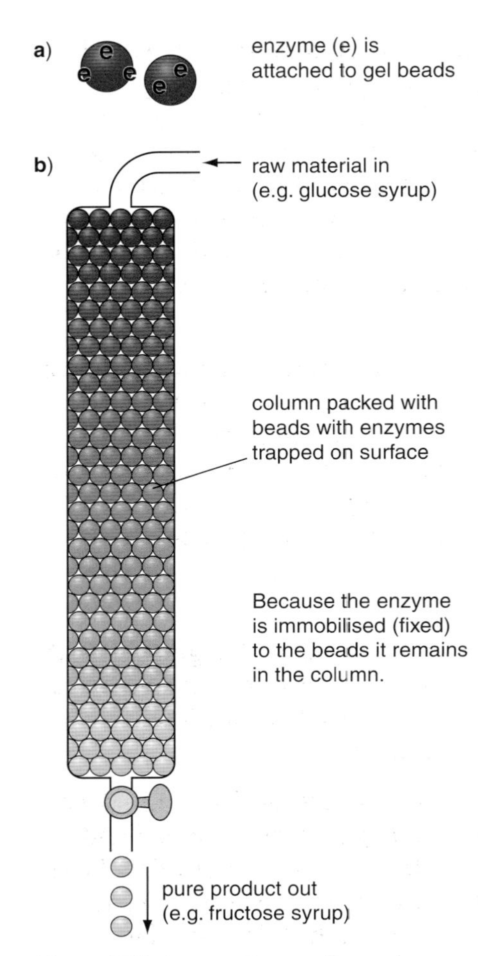

Figure 7.10 In a continuous flow column the substrate (glucose) is added through the top and reacts as it flows slowly over beads, which have the enzyme attached to the surface. The product can be collected from the bottom of the column.

20. List the economic advantages of using enzymes in the food industry.

21. a) In a continuous flow column, what is the advantage of having the enzymes immobilised on beads (Figure 7.10)?
    b) In a batch process, the raw materials and the enzymes are reacted together in a reactor vessel until an economic concentration of products has been formed. The remaining raw materials and the enzymes are then separated from the products. What are the advantages of 'continuous-flow' rather than 'batch' processes?

22. Complete the following table showing the use of enzymes in common food production.

Food product	Enzyme used in manufacture	Substrate	Product	Advantage of the product
Baby food	Protease		Amino acid	Readily available amino acids for baby growth
Slimming products	Invertase	Glucose syrup		Sweetens without adding calories
Soft drinks	Carbohydrase	(Maize) starch		
Baby food		Fats	Fatty acids	No digestion needed

Table 7.3

Other food products made from enzymes

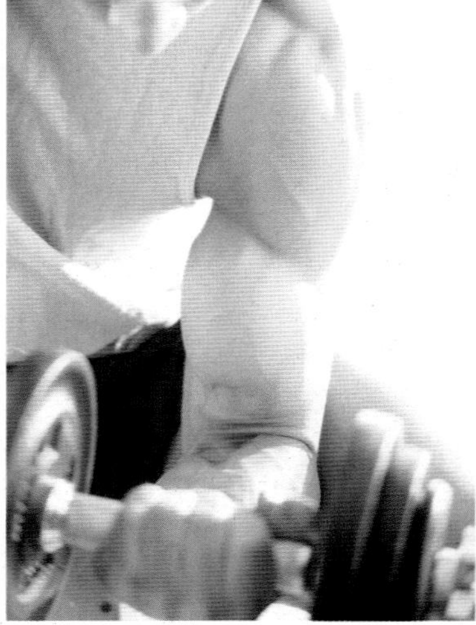

Figure 7.11 What is the connection between these two?

Figure 7.12 An enzyme product loved by many people

You would not expect much in common between food for babies and bodybuilders, but there is. Some baby food is pre-digested using proteases so that the baby can use the amino acids without digestion. Baby-milk powders are manufactured from cows' milk and are treated with enzymes to break down the proteins. This has the advantage that they are less likely to cause allergic reactions. The protein supplements taken by body builders are also pre-digested into amino acids, which can be quickly taken up for building and repairing muscles if consumed immediately after training.

Soft-centred chocolates start with the centre as a hard sugar paste, which can be easily handled for chocolate coating. The addition of invertase enzyme converts the sucrose to glucose and fructose syrup within 1 to 2 days. In other words, by the time the chocolates reach the shop the hard centre has become soft. This has other advantages in that fructose retains water within the centre and prevents the sugar crystallising out.

Laundry detergents

Washing used to be carried out at high temperatures with a great deal of agitation to break down dirt. This was costly in terms of electricity used to heat the water. Modern detergents use enzymes that function efficiently at low temperatures, reducing fuel costs.

Food or biological stains on clothes can include fats and oils from fried items, butter or oily dressings, proteins from eggs, grass or blood and starches from sauces made with flour. Biological washing powders have the following enzymes added: lipase to break down fats, proteases to digest proteins and carbohydrases to remove starch stains. Similar enzymes are used in dishwasher detergents.

Figure 7.13 What stains might be on the toddler's T-shirt? How could these be removed in a cool wash?

Although using 'biological detergents' improves stain removal and reduces costs, so much washing detergent is used in industrial production, in the hotel and catering industry and in the home that environmental factors must also be considered. Biological detergents are biodegradable, because enzymes are proteins and break down naturally in the environment. They also have other environmental advantages in energy efficiency and water efficiency.

23. Figure 7.13 shows a typical toddler's T-shirt. Explain how biological washing powder will solve the problem for each stain, including the egg, butter and tomato sauce.
24. What are the environmental advantages of using enzymes in detergents?
25. The instructions on the box state that biological washing powders are not effective when used at temperatures above 40 °C.
    a) Explain why this is so.
    b) If you had dropped fruit and cream down the front of your best silk shirt, what would be the advantage of using biological washing powder?

Activity – Stone-washed jeans

Figure 7.14

Denim fabric is made by weaving. The cross threads are unbleached cotton and the warp (vertical threads) are dyed with indigo. Indigo is a very stable dye and strong hypochlorite bleach (high pH and high temperature) is needed to remove it chemically.

Stone-washed jeans were originally produced literally with stones, in an enormous machine that shook 150 kg of pumice stone with 150 pairs of jeans for up to 6 hours. The stones damaged the cellulose fibres and released the indigo dye. This process was damaging to the seams of the jeans, the machines, the operators because of the pumice dust in the air, and it took a lot of water to rinse off all the grit particles from the jeans.

The modern method uses an enzyme called DeniMax, which produces the stone-wash effect faster without damage to the fabric and uses a minimum quantity of water.

The new enzyme process removes the need for pumice stones, the final look is the same and the jeans last longer.

1. List four problems with the old stone-wash system.
2. A new enzyme has been developed which is specific to indigo and which works under such mild conditions that it can also be used on stretch denim. The result of using the enzyme is to reduce the depth of the colour and produce an aged effect.
   What is meant by the term 'the enzyme is specific to indigo'?
3. What are the environmental advantages of the enzyme process used in the manufacturing of jeans?

What do the homemaker, laundry and hotel industry want from a detergent?	What are the environmental requirements for a detergent?
A good clean wash	No foam in water-treatment works
Low temperatures to reduce the cost of heating the water	Reduced amount of energy used, to reduce air pollution and greenhouse gas emissions from electricity generation
Less water used, as many house and businesses now have water meters	Reduced amount of water used, to avoid water shortages particularly in the south of England
Less agitation to reduce wear on fabrics	Biodegradable materials
All stains removed	
Fabrics to feel soft after washing	
Shorter washing time	

Table 7.4 The requirements for a good washing powder

Summary

- ✓ Enzymes are biological catalysts that enable reactions to take place at a lower temperature.
- ✓ Enzymes are **protein molecules** with a three-dimensional structure and are **denatured** by high temperatures.
- ✓ Enzymes function most effectively at an optimum temperature and optimum pH.
- ✓ Enzymes inside living cells catalyse respiration, photosynthesis and **protein synthesis**.
- ✓ **Aerobic respiration** releases energy from glucose. This energy is needed for movement, building new molecules and maintaining body temperature.
- ✓ Some enzymes are secreted from cells and function outside cells, for example in the digestive system
- ✓ Digestive enzymes catalyse the breakdown of food molecules into smaller molecules. **Amylase** breaks down starch into sugars, **proteases** break down proteins into amino acids, **lipases** break down lipids (fats and oils) into fatty acids and glycerol.
- ✓ Microorganisms that secrete enzymes outside their cells are used in industry for enzyme production.
- ✓ Enzymes are used in industry because they save time, money and environmental pollution.
- ✓ Enzymes are used in the food industry to produce glucose syrup from starch, fructose syrup used in slimming products and pre-digested baby food.
- ✓ Enzymes are used in domestic detergents to improve stain removal by digesting protein and fat in foods.

EXAMQUESTIONS

1 Pectin in fruit holds the cellulose fibres in the cell wall together. The enzyme pectinase breaks down the pectin and cell walls releasing the juice.

A student did a laboratory experiment to extract apple juice. She grated the apple to form pulp and weighed out 50.0 g into five separate beakers. The pulp was warmed in water baths and then the pectinase added and left for 15 minutes to react. Juice was collected by straining through fine mesh lining a filter funnel into a measuring cylinder.

The results are shown in Table 7.5.

a) Plot the results on a graph. *(3 marks)*

b) From your graph, suggest the optimum temperature for this enzyme. *(1 mark)*

c) Describe and explain the trend shown by the results. *(3 marks)*

Water bath temperature in °C	Volume of juice collected in cm^3
10	12
25	25
35	50
45	78
60	45
70	12
85	12

Table 7.5

d) Do you think there are enough results to make an accurate prediction of the volume of juice you would expect at 50 °C? How could you improve the experiment? *(3 marks)*

EXAMQUESTIONS

❷ A student was investigating two protease enzymes labelled A and B. Table 7.6 shows a copy of his practical record.

Tube number	Boiled egg white solution in cm^3	Enzyme	Other solutions added ($2\ cm^3$)	Result after 10 minutes
1	5	A	Water	No change
2	5	A	Dilute HCl	Clear
3	5	A	$NaCO_3$	No change
4	5	B	Water	Less cloudy
5	5	B	Dilute HCl	No change
6	5	B	$NaCO_3$	Clear

Table 7.6

a) Where in the human intestine would:
   i) enzyme A be produced *(1 mark)*
   ii) enzyme B be produced *(1 mark)*
   iii) enzyme B react? *(1 mark)*

b) What is the role of the hydrochloric acid? *(1 mark)*

c) Explain why enzyme B did not break down the egg white when dilute hydrochloric acid was present. *(2 marks)*

d) What substances in the intestine produce the optimum conditions for enzyme B to react? *(2 marks)*

❸ Quorn™ is produced in a continuous flow process in a 150 000 dm^3 reactor vessel. The filamentous fungus *Fusarium* is cultured in a medium based on glucose, ammonium compounds and vitamin B (biotin). Cultures are maintained at 28–30 °C and pH 6.0.

a) Examine the diagram. Is this an aerobic or anaerobic process? *(1 mark)*

b) What gas is extracted from the top of the vessel? *(1 mark)*

c) Why are ammonium compounds added? *(1 mark)*

d) What is meant by a continuous flow process? *(1 mark)*

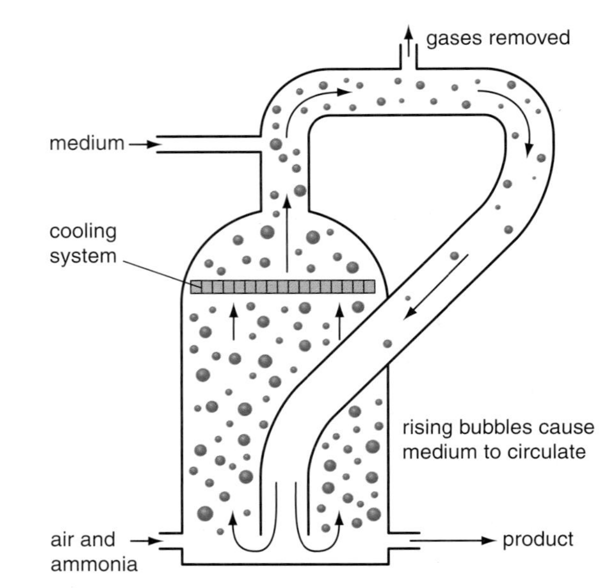

Figure 7.15

e) Biotin is a B group vitamin. Suggest how the fungus might use this. *(1 mark)*

f) Why do you think the temperature is maintained closely between 28 and 30 °C? *(1 mark)*

g) What do you predict would happen to the production rate if the system became more acidic? *(1 mark)*

❹ A student decided to carry out investigations with a low temperature biological washing powder, which had a banner on the box stating 'suitable for delicate coloureds – no bleach added'. He produced a standard dirty piece of material with egg yolk and mayonnaise stains on white cotton. He investigated a range of temperatures between 5 °C and 75 °C at 10-degree intervals. He found that the washed material was whitest in the range 25 °C to 35 °C.

a) Name two types of enzymes you would expect to find in biological washing powder. *(2 marks)*

b) Why was the result for the higher temperatures not so good? *(1 mark)*

c) Why do you think that the result at 5 °C was rather poor? *(2 marks)*

d) What are the environmental advantages of using a low temperature biological washing powder? *(2 marks)*

e) The instructions on the packet stated that gloves should be worn if clothes were hand-washed. Suggest why. *(1 mark)*

Chapter 8
How do our bodies control internal conditions and pass on characteristics?

At the end of this chapter you should:

- ✓ understand how the water and ion content of the body are controlled;
- ✓ know that ADH secreted by the pituitary gland controls the water content of urine;
- ✓ be able to explain how body temperature is maintained;
- ✓ know that the pancreas controls blood glucose concentration through the production of insulin;
- ✓ be able to explain how diabetes affects the body and how it can be treated;
- ✓ know how carbon dioxide and urea are removed from the body;
- ✓ understand the importance of DNA in living organisms;
- ✓ understand that mitosis is cell division that produces genetically identical cells;
- ✓ understand that meiosis is cell division that produces sex cells which are not genetically identical;
- ✓ be able to describe how a person's sex is determined;
- ✓ be able to explain how simple characteristics and some diseases are inherited;
- ✓ be able to complete genetic diagrams to illustrate inheritance;
- ✓ be able to make judgements about the applications of embryo screening.

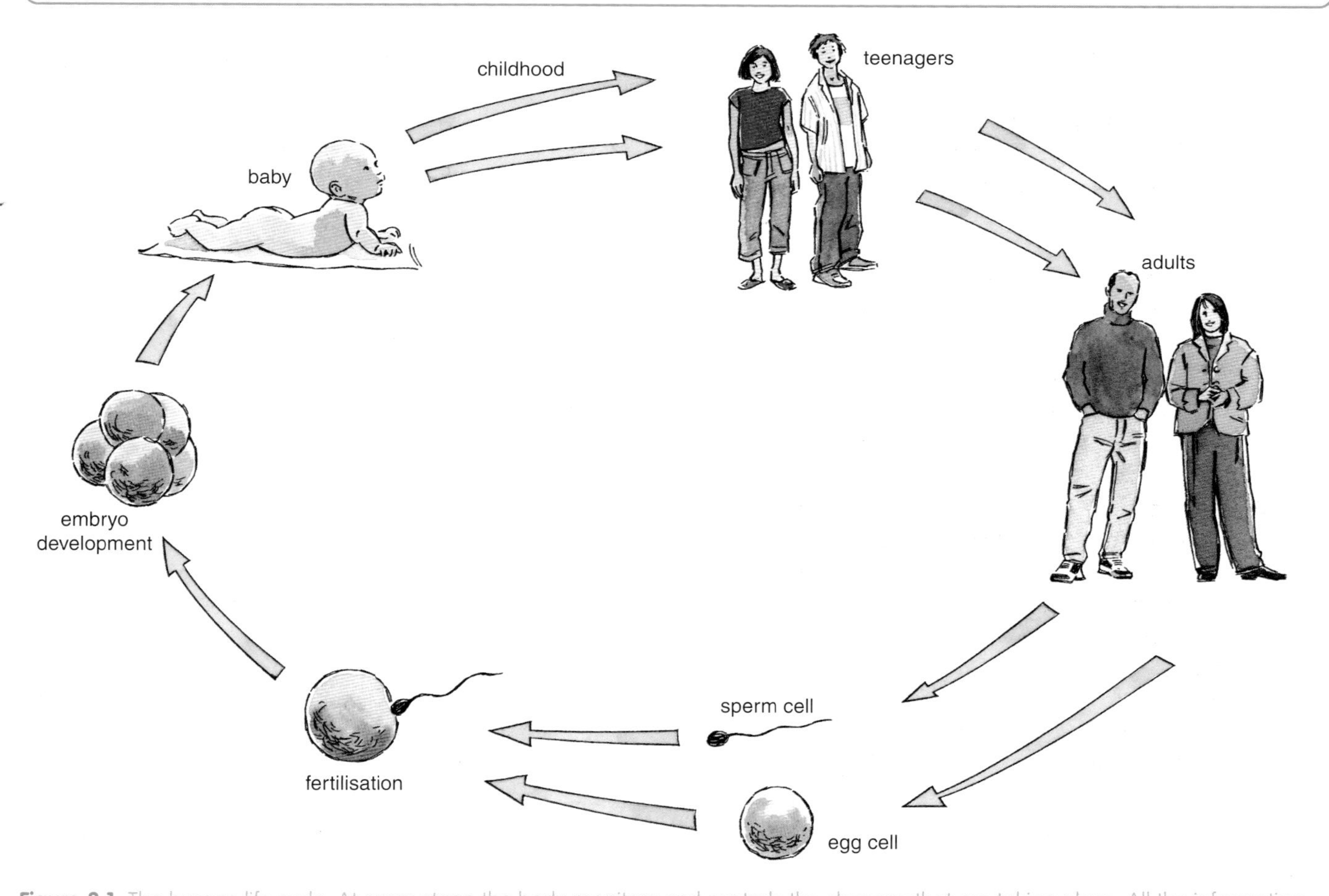

Figure 8.1 The human life cycle. At every stage the body monitors and controls the changes that are taking place. All the information that allows you to do this is inherited from your parents and can be passed onto your children. How can all this develop from just one cell?

How does your body control its internal conditions?

Homeostasis is the maintenance of steady conditions within the body, including water content, blood glucose concentration and body temperature.

Your body is constantly monitoring its internal conditions and making changes to try and control them. This means the body is controlling conditions in the cells, blood and body tissue. You will have been introduced to this idea whilst studying Chapter 1 when you learnt how hormones control the water and blood sugar levels in the body. The internal conditions which are controlled include the water and ion content of the body. Your temperature, oxygen, carbon dioxide and blood sugar levels are also controlled. **Homeostasis** is the name given to mechanisms in the body that regulate its internal conditions.

Figure 8.2 The girl is sweating to control her body temperature, which leads to water loss from the body. What else is lost from her body, in sweat?

Controlling the water and ion content of the body

When you get too hot you sweat to reduce your body temperature. The water in the sweat evaporates from your skin, cooling you down. But sweat doesn't only contain water. Ions, such as sodium and potassium, which are dissolved in your sweat, are also lost. Both water and ions are essential for health and we can obtain both when we eat and drink at mealtimes. You will remember the effects of dehydration and over-hydration from Chapter 5. Too much water in the blood can cause body cells to swell up or even burst. Too little water causes cells to shrivel. Too much or too little water in our cells means that they stop functioning properly. Body tissues, such as muscle, can stop working and cells can become permanently damaged.

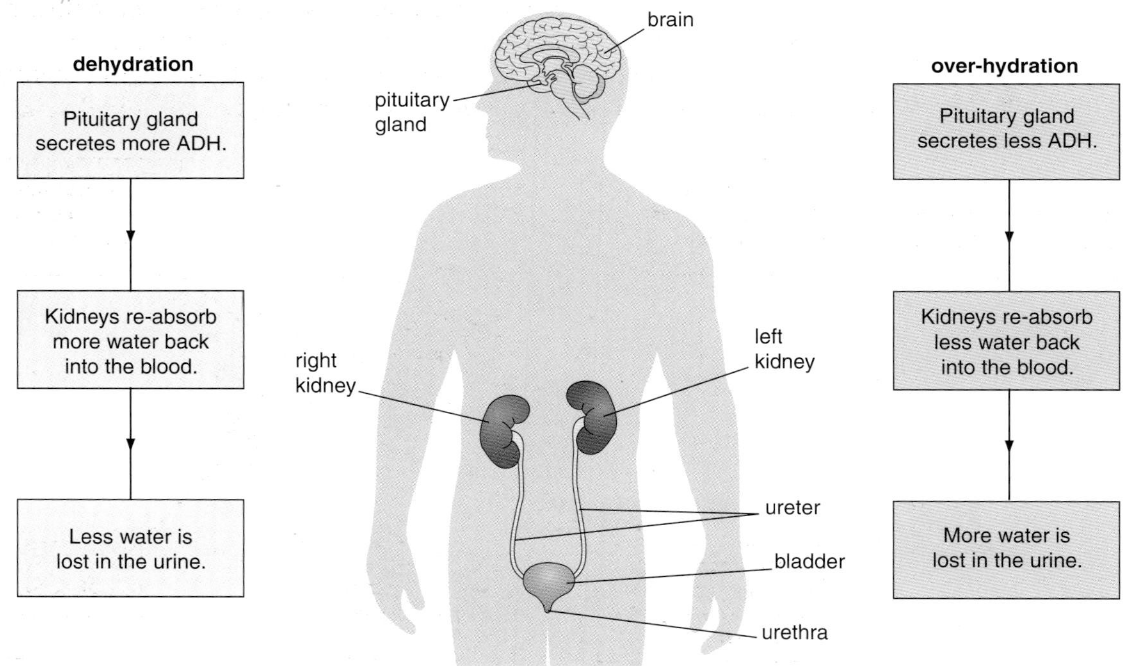

Figure 8.3 The pituitary gland and the kidney control the level of water in the blood. The kidney can adjust the amount of water in the urine.

1. Look at Figure 8.2 on page 126. The girl in the picture is losing water by sweating. How else may she lose water from her body?
2. She is also losing dissolved ions in her sweat. Explain how this will affect the water uptake by her cells. You may need to look back over the section on osmosis in Chapter 5.
3. How does sweating help her body to maintain the correct body temperature?
4. You can see that she is drinking to replace the water lost from her body. What would she have to do to replace the ions she has lost?
5. Imagine that the girl has only just started to dance.
   a) How would the pituitary gland respond when the girl initially starts losing water in her sweat?
   b) What effect would this have on the kidney?

The **thermoregulatory centre** in the brain monitors and controls body temperature.

In severe cases, dehydration or over-hydration can lead to unconsciousness or even death. Altering the concentration of ions dissolved in the blood also interferes with water uptake by the cells.

Controlling the water content of the blood is the job of the kidney. The kidney responds to a hormone called anti-diuretic hormone (ADH) which is secreted from the pituitary gland at the base of the brain. When a person becomes dehydrated, the pituitary gland secretes ADH which is transported in the blood to the kidneys. The kidneys respond by absorbing more water back into the blood and putting less water in urine. When a person becomes over-hydrated, less ADH is secreted, which causes the kidney to reabsorb less water into the blood. This results in more water going into urine and then being lost from the bladder when the person goes to the toilet.

Controlling body temperature

Birds and mammals have the ability to control their own body temperature. Other animals, such as reptiles, rely on the sun to warm up their blood. Humans usually maintain a body temperature between 36.5 °C and 37.5 °C. You will remember from Chapter 1 that if body temperature drops below 30 °C, sleepiness can set in, followed by unconsciousness. This can lead to death if the condition is not treated. Fortunately the body has a very effective way of monitoring and controlling its temperature.

Body temperature is monitored by a part of the brain called the **thermoregulatory centre**. The centre receives impulses from receptors in the surrounding area that monitor the temperature of blood flowing through the brain. The centre also receives impulses from temperature receptors in the skin. In Chapter 1 you learnt how receptors send electrical impulses to the brain along neurones. If the thermoregulatory centre receives impulses about changes in temperature it will send

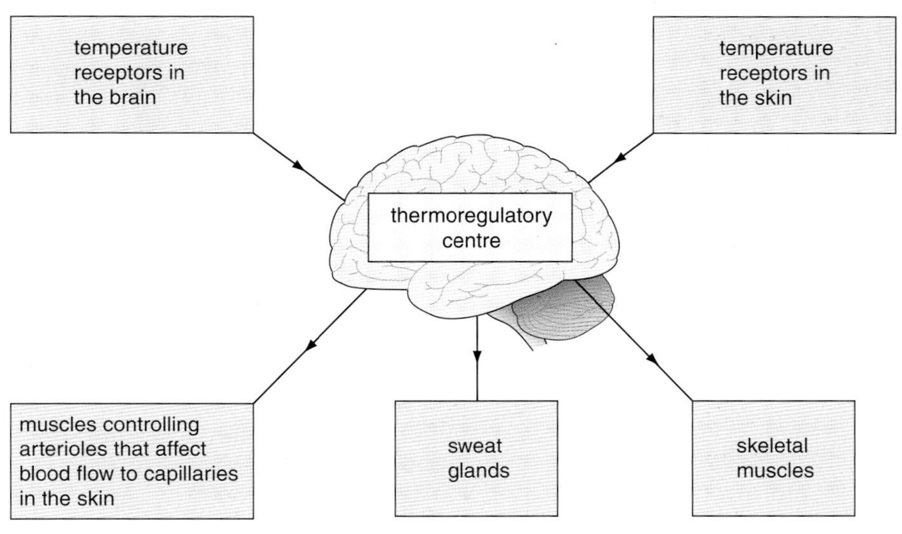

Figure 8.4 The thermoregulatory centre monitors and controls body temperature by co-ordinating a response to impulses received from receptors.

impulses out to other parts of the body. Various parts of the body make changes to adjust the temperature. All of these responses are automatic or reflex responses. Figure 8.4 summarises this system.

Sweat glands release sweat onto the surface of the skin, which evaporates to decrease body temperature.

So, how do these effectors change the body temperature? If the temperature in the body increases, the body can reduce the amount of heat it generates and increase the amount of heat it loses. If the temperature drops, the body can try to generate more heat and lose less heat. These changes are achieved by a combination of responses by the **sweat glands**, the skeletal muscles and muscles controlling blood flow through the arterioles to the skin capillaries. (Arterioles are small arteries that connect your main arteries to capillaries. Arteriole muscles relax and contract to increase or decrease blood flow to the skin capillaries to control body temperature. The blood vessels dilate (get wider).) Table 8.1 shows the three mechanisms for controlling body temperature.

	Increase in body temperature	Decrease in body temperature
Muscles controlling arterioles in the skin	Relax to allow more blood to flow into capillaries near the surface of the skin.	Contract to reduce the amount of blood flowing in capillaries near the surface of the skin.
Sweat glands	Secrete more sweat onto the surface of the skin.	Secrete less sweat on the surface of the skin.
Skeletal muscles	Do not cause shivering.	Contract and relax quickly to cause shivering.

Table 8.1 How the body responds to changes in temperature

Vasodilatation decreases body temperature by allowing more blood to flow through capillaries near the surface of the skin.

Vasoconstriction reduces the amount of blood flowing near the surface of the skin to retain heat inside the body.

Your skin plays a crucial role in controlling body temperature (Chapter 1). When body temperature increases, muscles that control blood flow through arterioles relax. This allows more blood to flow through the arterioles and into capillaries near the surface of the skin. Heat from the blood is then lost to the air surrounding the skin. This response is called **vasodilatation**. The opposite happens when body temperature drops. Blood flow near the surface of the skin is reduced so that less heat is lost. This response is called **vasoconstriction**.

Skin is also involved in the sweating response. When the body heats up sweat glands in your skin start to release sweat onto the skin's surface. Some of the heat near the surface of the skin causes the sweat to evaporate. As the particles in the sweat evaporate away from the skin they transfer heat energy with them, which causes the body temperature to drop. More water is lost from the body through sweating when it is hot than when it is cooler. We must always make sure that we drink enough water or eat enough food that contains a high percentage of water, such as fruit, to replace this water that has been lost.

When the body cools down, the muscles contract and relax rapidly – this is known as shivering. As the cells in the muscles respire they use up energy and release heat, which warms us up.

Temperature receptors in the brain detect a decrease in blood temperature.
Temperature receptors in the skin detect a decrease in skin temperature.
Electrical impulses are sent along neurones to the thermoregulatory centre of the brain.
The thermoregulatory centre sends impulses to skeletal and arteriole muscles.
Arteriole muscles contract to reduce blood flow to skin capillaries.
Skeletal muscles cause shivering.

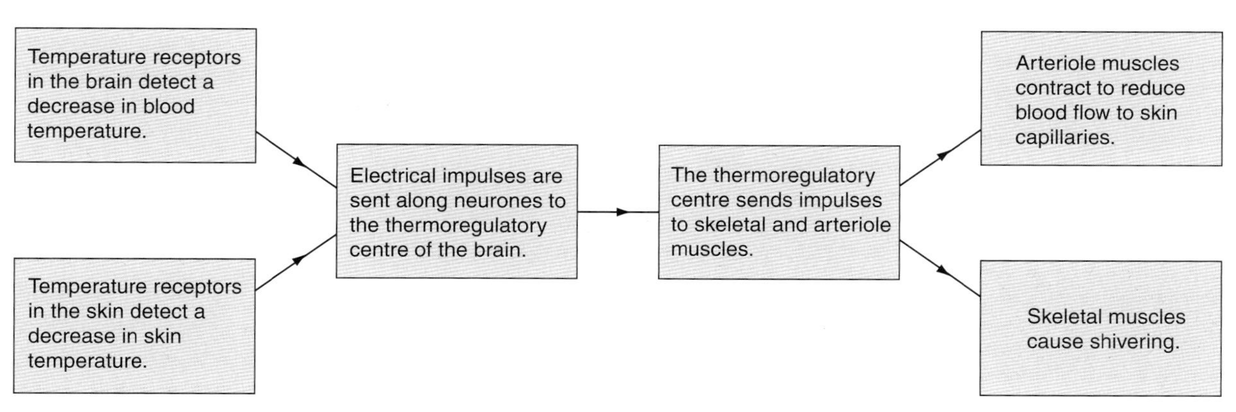

Figure 8.5 A flow diagram showing the body's response to a decrease in temperature

6. Figure 8.5 shows how the body responds to a decrease in temperature. Draw a similar flow diagram to show how the body responds to an increase in temperature.

7. Shivering is caused by your skeletal muscles contracting and relaxing repeatedly. Explain how this increases body temperature.

8. Look at Figure 8.6. It shows how the skin responds to changes in temperature.
   a) Which diagram shows the skin on a hot day?
   b) What changes can you see in this diagram that will decrease body temperature?
   c) Explain how each of these changes will decrease body temperature.

a)
pore
capillary loop
sweat duct
sweat gland
arteriole
Arteriole muscles contract to reduce blood flow through skin capillaries.
b)
pore releasing sweat onto the surface of the skin
dilated capillary loop
sweat duct carrying sweat to the surface of the skin
sweat gland producing sweat
arteriole
Arteriole muscles relax to allow more blood flow through skin capillaries.

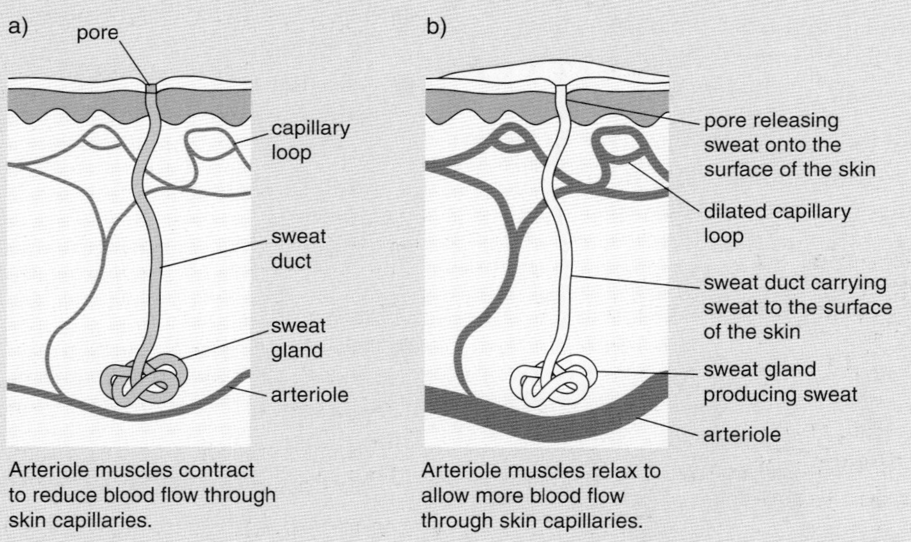

Figure 8.6 A cross-section of the skin on a) a cold and b) a hot day

Controlling blood sugar levels

The **pancreas** is an organ that monitors and controls blood glucose concentration. It produces the hormone insulin.

The **pancreas** is an organ that produces insulin. **Insulin** causes glucose to move out of the blood into cells. The production of insulin allows glucose to be stored in the liver and muscles and also to be used by body cells for respiration. Too much glucose in the blood can damage blood vessels and cells, so it is essential to have a mechanism to remove it. When blood glucose concentration is high, glucose is lost in the urine.

Insulin is a hormone that causes glucose to be moved from the blood into liver and muscle cells.

People who suffer from **diabetes** do not produce enough insulin in their pancreas, which leads to dangerously high blood glucose levels.

You will remember from Chapter 1 that insulin controls blood glucose levels. **Diabetes** is a disease that occurs when the pancreas does not produce enough insulin. This means that the amount of glucose in the blood rises to levels that are dangerously high and can be fatal. Before 1922, nothing was known about insulin and the effect it had on the body, so people with diabetes almost always died prematurely.

Fortunately, diabetes can now be treated because sufferers can inject insulin to reduce blood glucose levels. People with diabetes can also control their illness by being very careful about what foods they eat.

Activity – Discovering insulin

A doctor called Frederick Banting had studied diabetes and thought that the condition was linked to the pancreas. With an assistant called Charles Best, he carried out a set of experiments using dogs. Best first tied up the tube connecting the dog's pancreas to its blood system. He then measured the dog's blood glucose levels and noted that these rose dramatically over an eight-week period. He deduced from his observations that the pancreas must produce something that controlled the amount of glucose in the blood.

He then removed the pancreas of the dog, along with the pancreas from another dog. He ground up the pancreas from the second dog with salt water, and injected this mixture into the first dog. Best observed that the dog's blood glucose levels were temporarily reduced. However the levels soon rose and the dog died the following day. Banting used these findings to conclude that the pancreas produced a substance that controlled blood glucose level. He called this substance insulin. Banting went on to show that insulin obtained from healthy dogs could be used to treat diabetic dogs and won the Nobel Prize for his work.

Figure 8.7 Frederick Banting and Charles Best with the dog they used in their experiment

1. Draw a flow diagram summarising the work of Banting and Best.
2. The first set of experiments were carried out using just two dogs. Bearing this in mind, do you think Banting was right to be so confident of his conclusion? Give reasons for your answer.
3. Suggest some improvements that Banting could have made to his experiment to collect more reliable results.
4. People often refer to Banting's work when they argue for the use of animals in experiments to develop new medical treatments. Do you think that referring back to experiments carried out in 1922 is a strong enough argument to justify animal experiments today? Explain your opinion.

Once people realised that Banting's discovery could lead to a treatment for diabetes, further research into insulin was quickly carried out. It was found that insulin could be extracted from the pancreases of cows and pigs and used successfully to treat diabetes. Many thousands of lives were saved in this way but a huge amount of insulin was needed, which meant that lots of pancreases had to be collected from abattoirs.

Nowadays insulin is manufactured using genetically modified bacteria. This removes the need to use the pancreases from slaughtered animals.

5 Some people objected to using insulin from animals' pancreases. What reasons do you think they had for this viewpoint?

6 Produce an information sheet that could be included on a website or in a leaflet aimed at people who have just been diagnosed with diabetes. You should include a summary describing the disease and information about the different ways of treating it.

A search for diabetes at www.bbc.co.uk is an excellent starting point. You may find that www.diabetes.org.uk is also useful.

How does your body get rid of waste products?

Some of the chemical reactions that take place in your body produce useful products such as proteins. Respiration is essential because it releases energy from glucose. However, these reactions also produce waste products that can be toxic to the body. Fortunately your body has ways of removing these waste products so that they do not reach harmful levels in the blood.

How is carbon dioxide removed from the body?

Respiration is a series of chemical reactions that take place in cells. It is summarised by the equation:

glucose + oxygen → carbon dioxide + water + energy

The carbon dioxide produced moves from the cells into the blood by diffusion (see Chapter 5). Too much carbon dioxide in the blood can be dangerous because it lowers the blood pH and inhibits enzyme action. When the blood flows through the lungs, carbon dioxide diffuses out of the blood and into the air in the lungs. At the same time oxygen diffuses from the air into the blood. This takes place across the walls of tiny structures called alveoli and is called gas exchange. Once the carbon dioxide is in the air in the lungs it is breathed out of the body.

How are other types of waste removed from the body?

Proteins are digested by enzymes, called proteases, and broken down into amino acids (enzymes are described in Section 7.1). These amino acids are used by the body to make other useful proteins that are needed for growth and to control body processes. You often have more amino acids in your blood than your body can use. The liver converts these excess amino acids into a chemical called urea. Urea can harm the body, so it must be removed. Figure 8.8 shows how this happens.

9. Explain why oxygen diffuses from the alveoli in the lungs to the blood, and why carbon dioxide diffuses from the blood into the alveoli.
10. Why does your breathing rate increase when you go for a run? Refer to oxygen and carbon dioxide in your answer.
11. What is the difference between urea and urine?
12. People who suffer from liver damage are sometimes asked to reduce the amount of protein in their diet. Explain the scientific reason behind this advice.
13. Urine can be tested for the presence of amino acids. What possible medical problems do you think somebody might have if amino acids are detected in their urine?

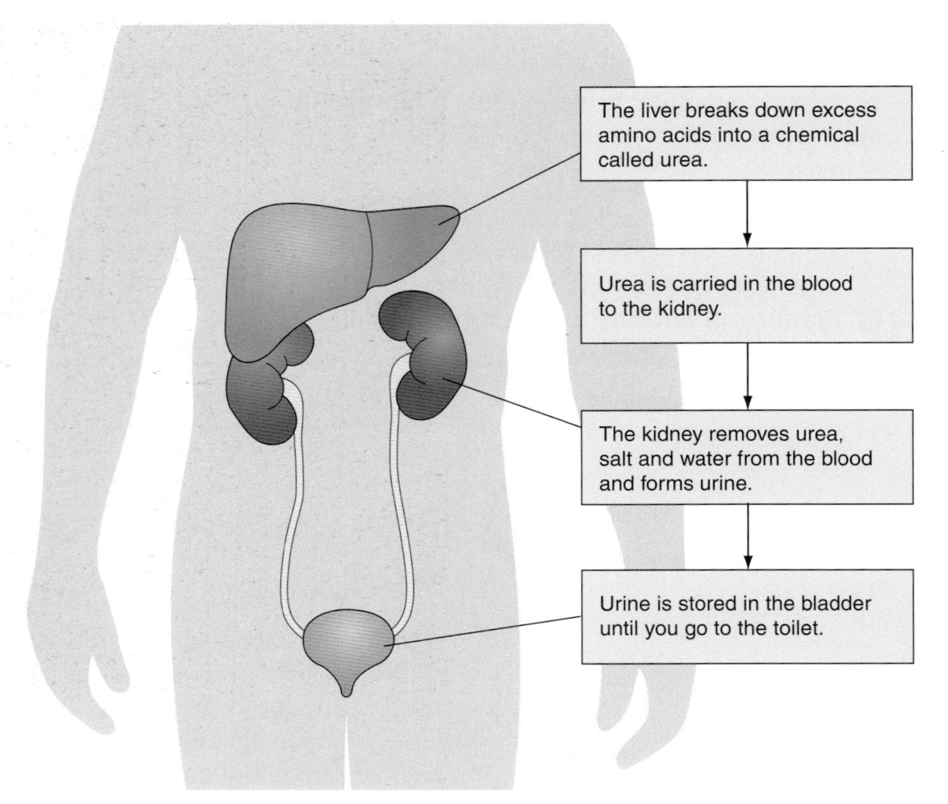

Figure 8.8 How excess amino acids are removed from the body

As you can see in Figure 8.8, the liver and the kidney work together to remove amino acids from the body. The liver breaks down excess amino acids by a series of chemical reactions. This produces urea, which dissolves in the blood plasma and is then transported to the kidneys. Since it is dissolved in the plasma it is removed from the blood along with some of the water it is dissolved in. This mixture, along with salt, makes up the urine that is stored in the bladder until you go to the toilet.

8.3 Why is cell division so important?

You will remember from Chapter 4 that chromosomes are structures found in the nucleus of cells. When cells divide they form new cells with copies of these chromosomes. Without cell division people would not grow, reproduce or be able to repair damaged body tissue.

Chromosomes are molecules of DNA found in cells. A single chromosome is a single strand of DNA.

DNA (deoxyribonucleic acid) is a long-chain molecule, made up from a series of bases.

What are chromosomes and genes?

In humans, there are 23 pairs of chromosomes inside the nucleus of a normal cell, or 46 in total. This number is different in other species. **Chromosomes** are each made up of a long molecule of **DNA** (deoxyribonucleic acid). A single chromosome is a molecule of DNA about 5 cm long. This is amazing when you consider that chromosomes

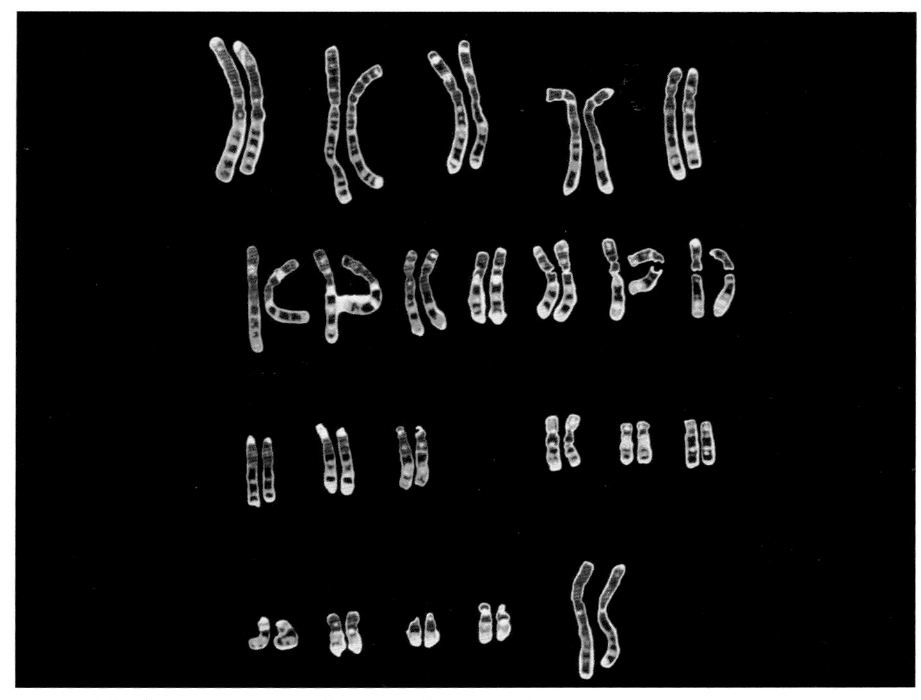

Figure 8.9 The chromosomes from one human cell

can be seen only under very powerful microscopes. To achieve this each DNA molecule is very thin and has to be coiled up very tightly to form a chromosome.

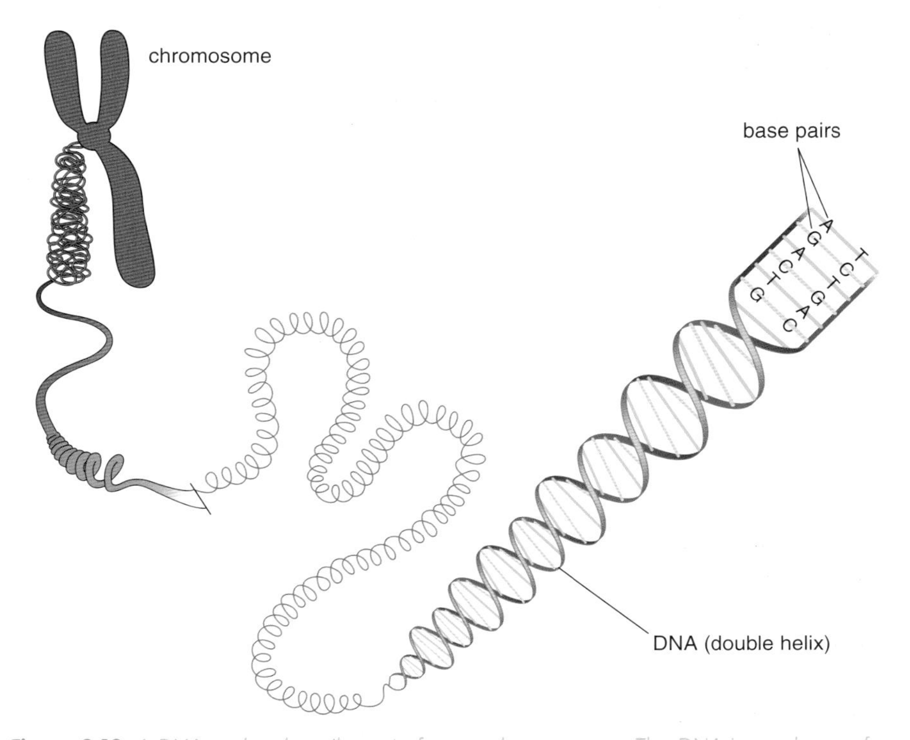

Figure 8.10 A DNA molecule coils up to form a chromosome. The DNA is made up of a series of bases: A, T, G and C. Varying combinations of these bases make up the instructions for producing proteins.

A **gene** is a length of DNA that codes for a specific protein.

DNA fingerprinting is a technique that uses the non-coding pieces of our DNA to provide a unique identification sequence.

14 The DNA in one chromosome is far longer than the chromosome itself. Why is this? Think about how the shortening of DNA to a chromosome is achieved.

15 a) Why do there have to be so many genes in a human cell?
b) Sometimes a gene mutates. This means that the sequence of bases on that gene is altered. What effect will this have on the protein that the gene is responsible for making? Explain your answer.

A **gene** is a small section of a DNA molecule. There are between 20 000 and 25 000 genes in each human nucleus. You can see in Figure 8.10 that a DNA molecule contains a series of bases that are represented by the letters G, C, A and T. Bases are molecules that form part of DNA. A gene is therefore really a sequence of these bases. The sequence of bases in a gene acts as a code (instructions) for making a protein.

All humans have the same types of genes, so what makes us unique? Some areas of DNA, called non-coding sequences, don't contain any genes. These areas are different for each person in the world, unless you have an identical twin. This means that a person can be identified from the non-coding DNA obtained from a swab of their cells. This technique, called **DNA fingerprinting**, can be used to link people to a crime scene. If, for example, skin cells are found at a crime scene, then the DNA inside them can be removed and identified. If the non-coding DNA matches the DNA collected from a suspect then it places the suspect at the scene.

Section 8.2 referred to amino acids that are obtained when proteins are digested. A gene's job is to take these amino acids and put them together in a specific order to make new proteins that the body needs. This order of amino acids is controlled by the order of the DNA bases. Since each gene is a unique sequence of bases, each gene puts the amino acids in a different order and makes a different protein. This takes place in a ribosome (see Chapter 5). The proteins made can be enzymes or hormones, or they may also be used to form new body tissue, such as muscle, or even to produce the colour in your eyes.

What happens to chromosomes when cells divide?

A cell divides by **mitosis** to produce two genetically identical cells with a full set of chromosomes.

You will remember from Chapter 5 that a cell can make a genetically identical copy of itself. This takes place when you grow and also when cells need to be replaced if body tissue becomes damaged. For example, if you were to cut yourself, the new skin that would form over the cut would be made by skin cells dividing to make new skin cells. When a cell divides in this way it forms two cells that are both genetically identical to the original cell. This can happen because chromosomes have the amazing ability to make identical copies of themselves. This type of cell division that produces genetically identical cells is called **mitosis**. Figure 8.11 opposite shows how it takes place.

Mitosis also takes place in some plants and in single-celled organisms such as bacteria and yeast during asexual reproduction. Offspring produced by this means have exactly the same genes as their parents, because they contain an exact copy of their chromosomes. They are often referred to as clones.

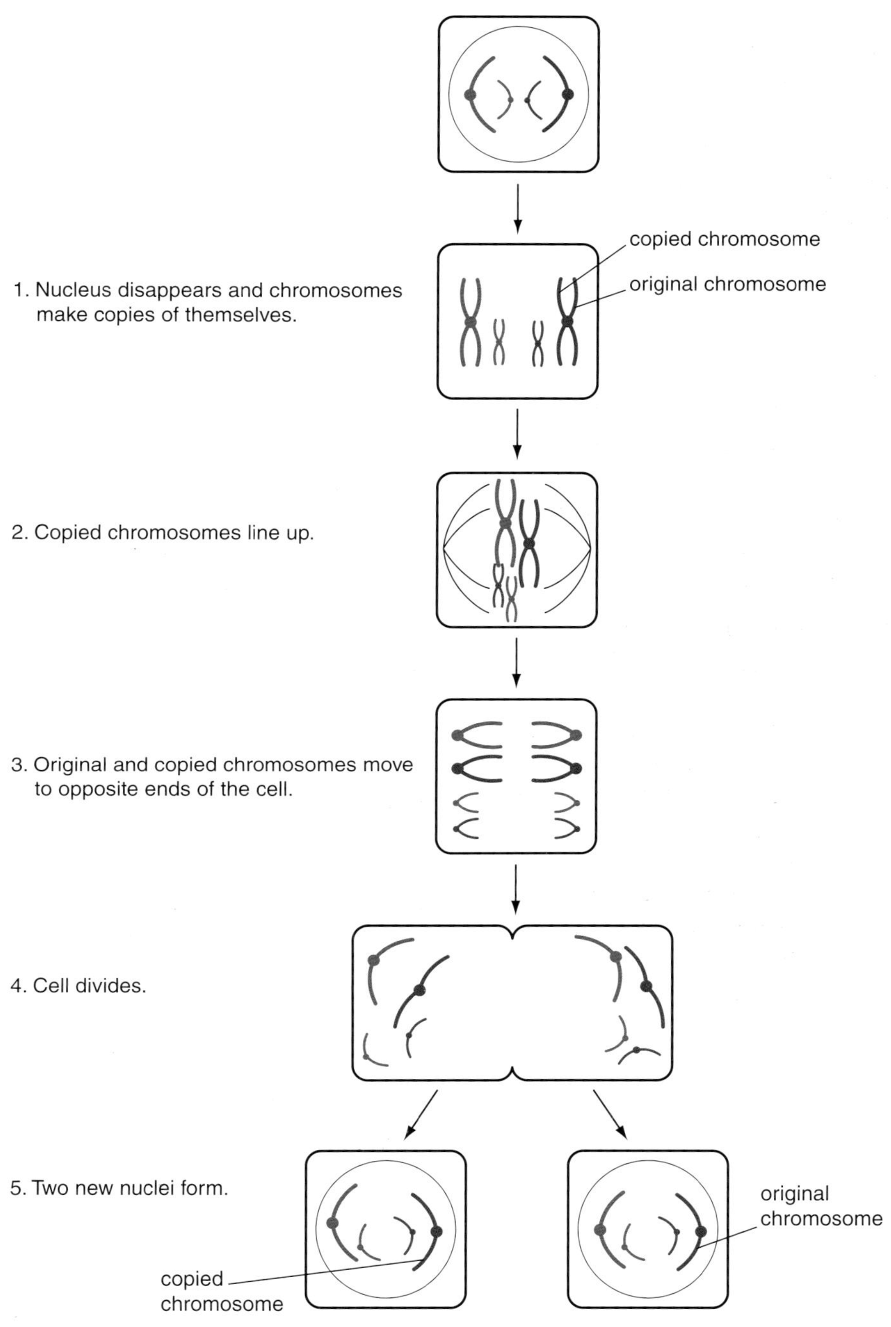

Figure 8.11 The diagram shows the main stages in mitosis for a cell with just two pairs of chromosomes. The original and the copied chromosomes are labelled so you can see that the new cell is an exact copy of the original cell.

Gametes are sex cells. These include sperm cells, egg cells and pollen cells.

A cell divides by **meiosis** to produce four genetically non-identical cells, each with half the original number of chromosomes. Meiosis forms cells called gametes.

A second type of cell division, found in most animals and plants, has only one function – to make sex cells. In animals the sex cells, or **gametes**, are sperm cells and egg cells. In plants the male gametes are pollen cells. In humans, this type of cell division only takes place in the reproductive organs, the testes and ovaries. Cell division that forms gametes is called **meiosis**. The cells that are formed by meiosis are not

16 What are the differences between mitosis and meiosis?

17 Look back at Figure 8.1 on page 125, which shows a human life cycle. At what stages in the life cycle would you find a) mitosis and b) meiosis taking place?

18 Why do gametes have only half the number of chromosomes found in a normal body cell?

19 It is almost impossible for two human egg cells produced from one person to be identical. Explain why this is the case.

identical to the original cell. They only have half the number of chromosomes. This means that in humans a gamete has 23 chromosomes. Figure 8.9 on page 133 shows a full set of human chromosomes. A gamete will only contain one chromosome from each pair.

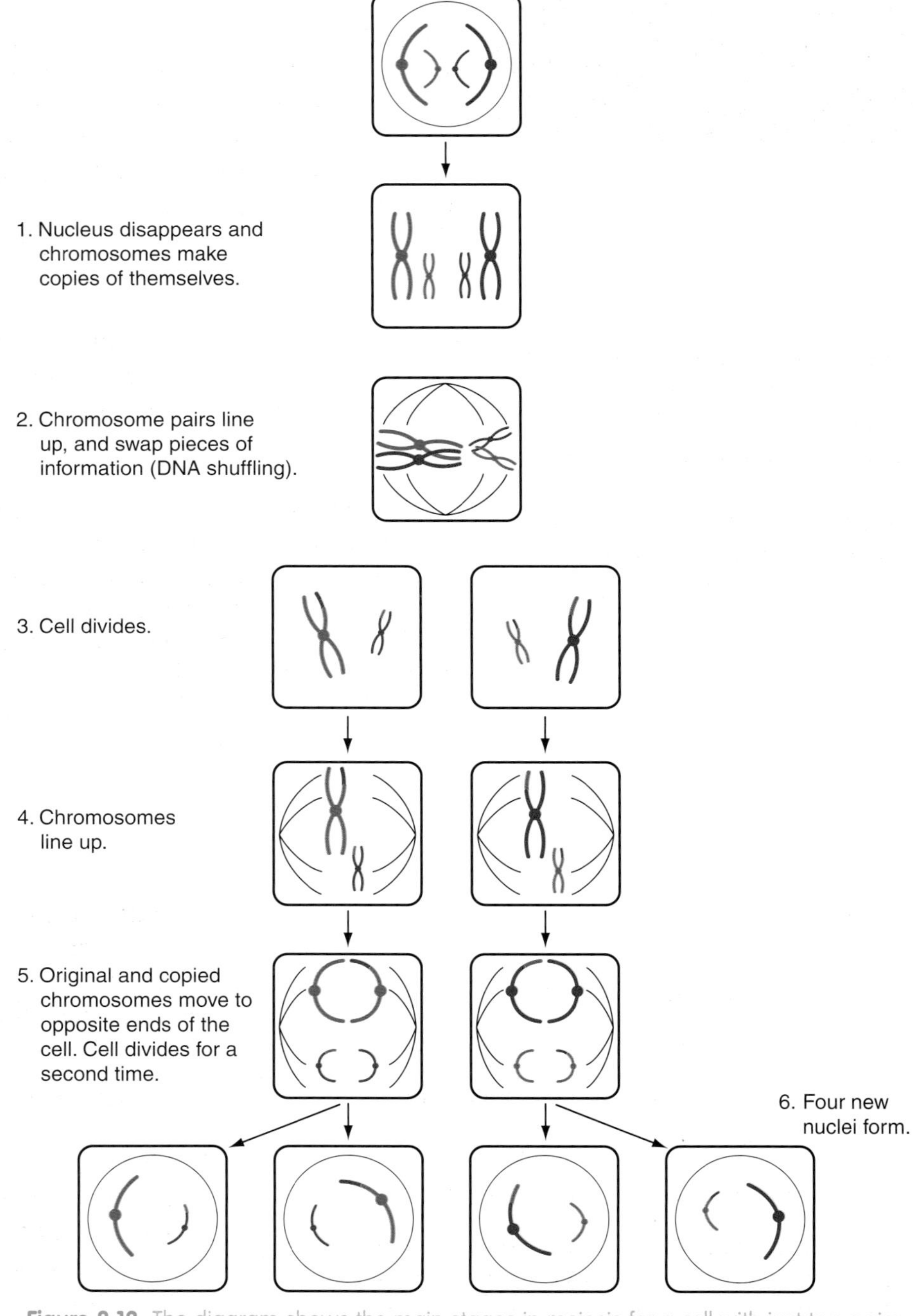

Figure 8.12 The diagram shows the main stages in meiosis for a cell with just two pairs of chromosomes. Once the matching chromosomes have paired up, they swap pieces of information between them. This is called DNA shuffling. The cell then divides twice and ends up with four cells (gametes), each with half the original number of chromosomes.

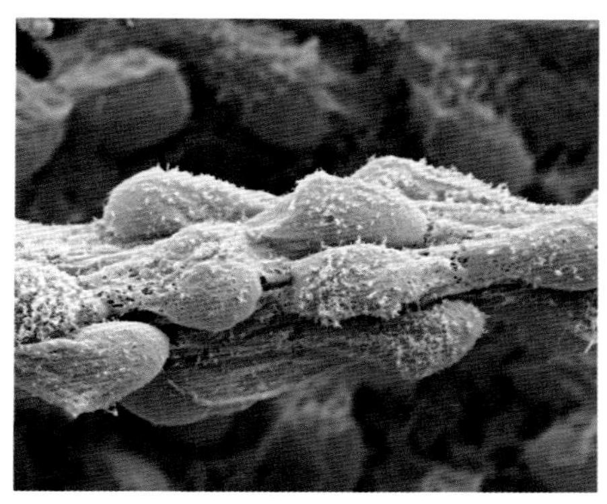

Figure 8.13 Embryonic stem cells, like those shown here, can differentiate into completely different types of cell. Do these cells have any specialised features?

Differentiation is the process by which an unspecialised early embryonic cell acquires the features of a specialised cell.

Undifferentiated cells have no specialised features. Stem cells are a type of undifferentiated cell.

Stem cells and specialised cells

Figure 8.13 shows some human stem cells. They look like normal cells, so why are they so important? Stem cells can carry out one function that other cells cannot. When stem cells divide they can make specialised cells.

Compare these with the nerve cells studied in Chapter 1. Nerve cells have a specialised structure that allows them to carry electrical impulses around the body. Chapter 5 describes some other specialised cells. Nerve cells and most types of animal cell **differentiate** at an early stage of embryo development and then divide to produce only that type of specialised cell. However, many plant cells retain the ability to produce different differentiated cells throughout their life.

Stem cells are **undifferentiated**, which means that they have no specialised features to carry out a specific function. It is still not understood exactly how they do this but, under the right conditions, stem cells can develop into other types of cells such as liver, brain or heart muscle cells.

There are two groups of stem cells: embryonic stem cells and adult stem cells. Embryonic stem cells are found in the embryo soon after an egg cell has been fertilised. Embryonic stem cells differentiate into all the different specialised cells that eventually form a fetus and then a baby. Adult stem cells are found inside some body tissue, such as bone marrow. Adult stem cells make the type of specialised cells in the tissue where they are found. For example, stem cells in the pancreas produce pancreas cells that are specialised to make insulin.

Activity – Stem cells and medical research

'Stem cell promise lures patients'

'Stem cells treat blood disorder'

'Stem cell heart cure to be tested'

'MS sufferer in stem cell gamble'

'Winston warns of stem cell "hype" '

These are just a few recent headlines from articles about research that focuses on human stem cells. Of all the current biomedical research, stem cell research has probably attracted more attention than any other from the media. But why is this?

To understand the level of interest and range of views expressed, it is crucial to know exactly what happens when stem cells divide (see above) and the potential medical benefits these cells could bring.

1. Explain why stem cells are essential in human development.
2. Figure 8.14 shows some of the specialised cells that stem cells produce. Each one has a specific structure that allows it to carry out its function. Find out and state the function of each of these specialised cells. Explain how each cell's structure enables it to carry out its function.

3. Stem cells can be collected and kept alive outside the body. With this in mind, suggest two possible uses scientists could make of stem cells. For each suggestion explain how you think scientists would carry out the procedure. Put the main points into a flow diagram.

Current stem cell research involves manipulating embryonic stem cells to try to produce specialised cells. Hypothetically, these new specialised cells could replace cells that have been damaged either by disease or by an accident. Currently only small amounts of embryonic stem cells can be created using this technique, and scientists have not yet been able to create specialised cells. However, scientists believe that one day it may be possible to grow entire organs using this process.

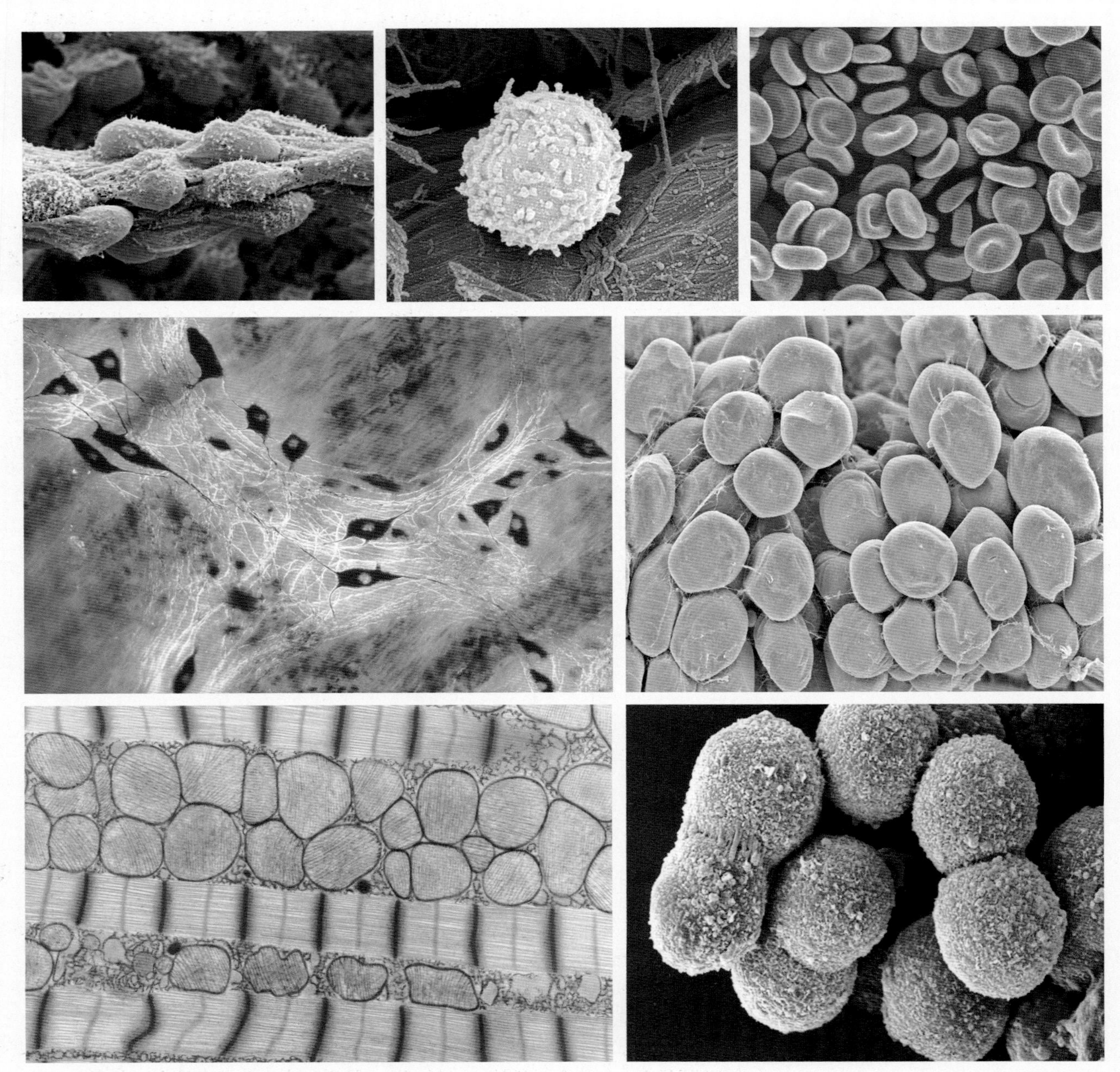

Figure 8.14 Some of the specialised cells that can develop from stem cells

Stem cell scientists take an unfertilised egg cell from a donor and insert the patient's DNA into its nucleus. The cell then divides in a laboratory until a clump of cells is formed. This clump of cells is effectively a tiny embryo, which if implanted into the womb of a woman could develop into a baby. Like all embryos, it contains embryonic stem cells, which are removed and then manipulated to try to produce the specialised cells required.

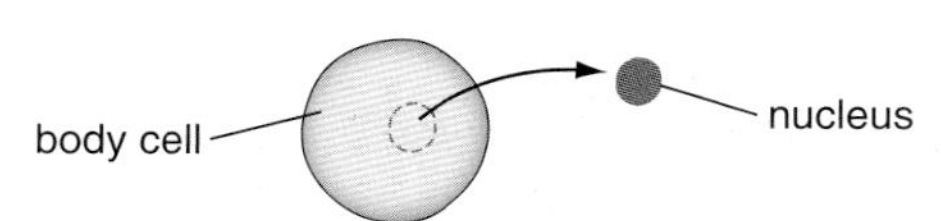

1. A nucleus is removed from a body cell of the patient.

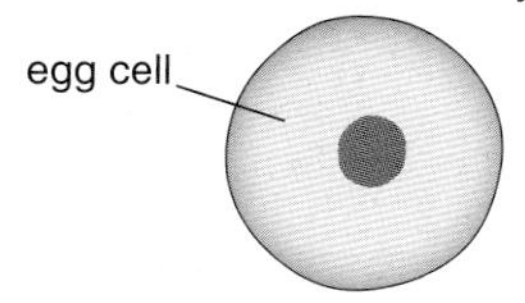

2. An egg cell is removed from an adult female.

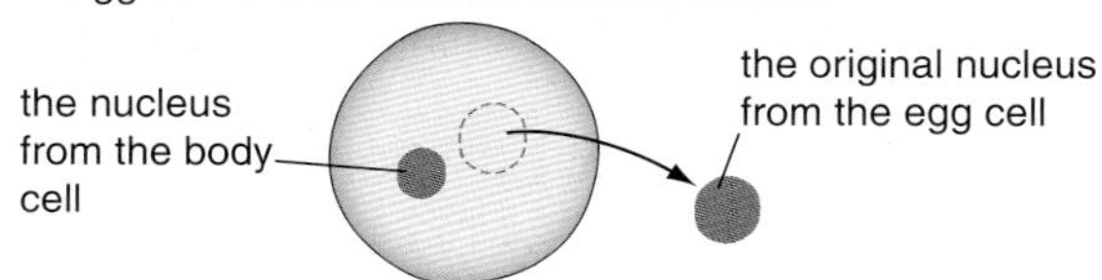

3. The nucleus is removed from the egg cell and replaced with the nucleus from the first adult body cell.

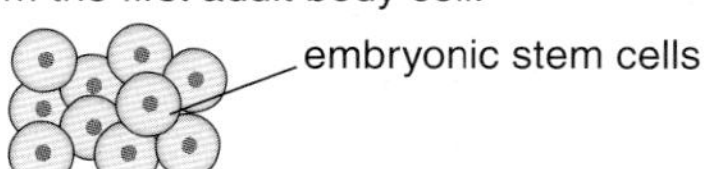

4. The egg cell starts to develop into embryonic stem cells.

Figure 8.15 The procedure used to produce stem cells that match the DNA of a patient. This means the new cells or tissue produced will not be rejected by the patient's body.

Stem cells produced like this may one day be used to treat diseases such as Parkinson's disease. Patients with Parkinson's disease have nerve cells that are damaged. As more and more nerve cells become damaged, the patient suffers from progressive paralysis. Stem cells could be manipulated in the laboratory to produce healthy nerve cells to replace those that are damaged in the Parkinson's patient.

Opinions on stem cell research are mixed. Some people support stem cell research as they feel it could eventually give patients with previously incurable diseases a better quality of life. Other people argue against the use of stem cell research. They insist that researchers should not be allowed to experiment on embryos since these balls of cells have the capacity to develop into a complete human being.

4. In small groups, debate reasons that people may have to support or oppose embryonic stem cell research. Write these down as a list of reasons for and a list against the research.
5. Now put the reasons in each list into order based upon which ones present the strongest arguments for and against stem cell research.
6. Finally, write a paragraph expressing your views about this issue. Support your views with reasons.

How are characteristics passed on from one generation to the next?

When a sperm cell fertilises an egg cell the genetic information from the mother and father is brought together. Fertilisation produces a cell, called a zygote, that contains a full set of chromosomes. Each gamete (one from the mother, one from the father) carries one chromosome from each pair found in a normal human body cell, so when the two gametes come together the full set of 23 pairs is produced. At fertilisation, all the characteristics that are controlled by your genes are determined, for example, your eye colour and hair colour. After fertilisation, the zygote divides by mitosis and develops into an embryo.

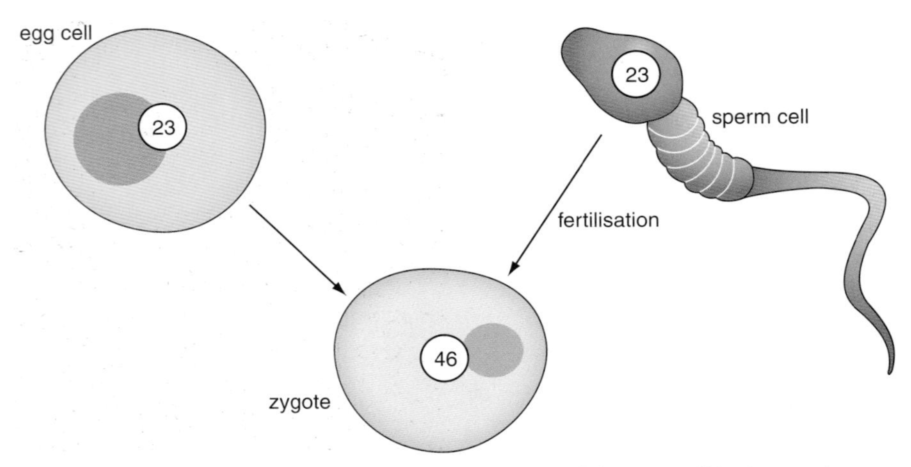

Figure 8.16 During fertilisation in humans the sperm and the egg cell bring together 23 chromosomes from each parent to form the first cell that goes on to produce a new human being. After fertilisation, this cell has the full number of chromosomes – 46.

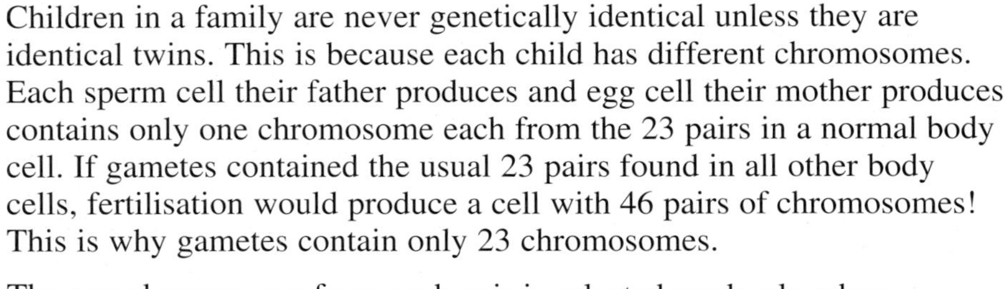

Children in a family are never genetically identical unless they are identical twins. This is because each child has different chromosomes. Each sperm cell their father produces and egg cell their mother produces contains only one chromosome each from the 23 pairs in a normal body cell. If gametes contained the usual 23 pairs found in all other body cells, fertilisation would produce a cell with 46 pairs of chromosomes! This is why gametes contain only 23 chromosomes.

Figure 8.17 Each child in this family inherited half their chromosomes from their mother and half from their father. The differences between them exist because each gamete carries different chromosomes.

The one chromosome from each pair is selected randomly when gametes are formed by meiosis. If you consider that there are 23 pairs, it is incredibly unlikely that any two gametes will have exactly the same 23 chromosomes. Imagine you had two sets of 23 numbered cards. The 46 cards represent the 23 pairs of chromosomes in a normal human cell. The two sets are numbered 1–23 but one set is coloured red and the other blue. The red set represent 23 chromosomes (one from each chromosome pair) and the blue set represents the remaining 23 chromosomes from each pair. If you were to randomly pick one card from each of the 23 pairs, and then do it again, there is only a tiny chance that you would end up with the same 23 cards both times. This is a bit like the way in which random chromosomes are selected from each chromosome pair, when gametes are produced by meiosis.

How is the sex of a person determined?

One of the most obvious differences between children in a family is their sex, whether they are a boy or a girl. A person's sex is determined at fertilisation because it is controlled by the genes found on a pair of chromosomes. All the differences between males and females are actually controlled by just one chromosome.

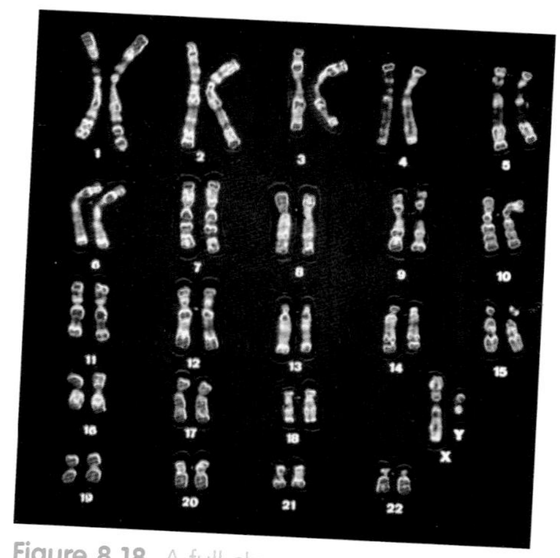

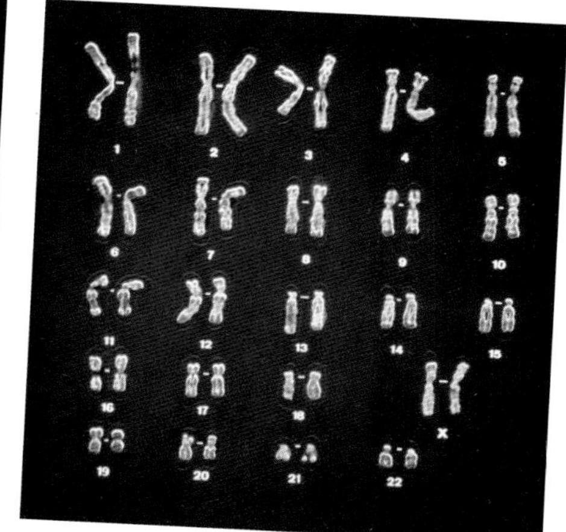

Figure 8.18 A full chromosome set from a man and a woman. Look at the last pair of chromosomes (marked X or Y) in each set.

The 23rd pair of chromosomes in each cell are the **sex chromosomes** and they determine a person's sex. XX = female; XY = male.

A **genetic diagram** shows all the possible combinations of chromosomes or genes that could be passed on during fertilisation.

You can see in Figure 8.18 that both chromosomes in the last pair of 23 are the same in females, but different in males. These chromosomes are called the **sex chromosomes** and are the 23rd pair in each cell. In females both of the sex chromosomes are called X chromosomes. Males have one X and one Y sex chromosome. Every egg cell that a woman produces carries an X chromosome. Half the sperm cells that a man produces carry an X and half carry a Y chromosome. Remember that each gamete gets one randomly selected chromosome from each pair during meiosis. So if an X-carrying sperm cell fertilises an egg, then a baby girl will be born. If a Y-carrying sperm cell fertilises an egg, then a boy will be born.

The **genetic diagram** in Figure 8.19 shows how sex chromosomes are passed on from parents to their children.

20 Explain how sexual reproduction leads to variation between children in the same family.

21 Why is it essential that gametes in humans are produced by meiosis and not mitosis?

22 'The sex of a baby is determined entirely by the father.' Do you consider this statement to be correct? Explain your answer by referring to how sex is inherited.

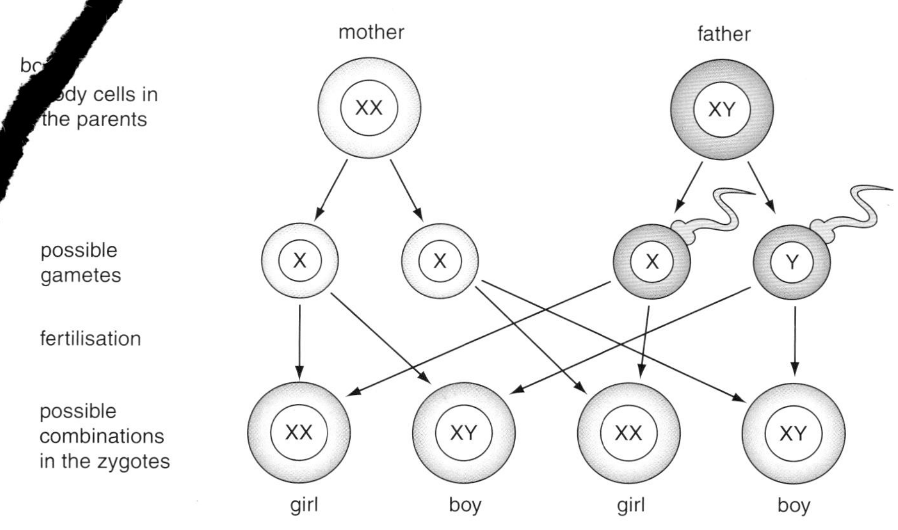

Figure 8.19 This genetic diagram shows how sex chromosomes are inherited.

> **23** There are now two methods available to parents which enable them to select the sex of their baby. The first is called the 'sperm cell selection method' and the second is known as the 'embryo selection method'. Both of these methods are based on being able to detect the X and Y chromosomes in cells.
> a) Explain how you think each of these methods works.
> b) Give two reasons for and two reasons against giving people the chance to select the sex of their baby.

How are some characteristics controlled by a single gene?

You will remember the history of ideas about inheritance from Chapter 4. An Austrian monk called Gregor Mendel did not agree with the explanation of inheritance held in the 1800s, known as the 'blending theory'. This suggested that all your characteristics were simply a mixture of your parents' characteristics. As Mendel was working in the monastery gardens he noticed that pea plants had either red or white flowers. When they were bred together an individual new plant still produced red or white flowers, but never pink or a mixture of red and white flowers. To him this made it clear that inherited characteristics were not just a mixture of parental characteristics.

To test this idea, he carried out many repeated experiments and collected reliable evidence. This evidence supported his idea that individual characteristics, such as flower colour, were inherited separately and were not simply a mixture of the parents' characteristics. However, his discovery was not widely accepted until after his death when scientists carried out similar experiments that led to the same conclusions. Mendel's reports were later translated into English, which meant that they had a far wider audience. This undoubtedly helped persuade more people that Mendel's idea was a far better explanation of inheritance than the blending theory.

> **Alleles** are alternative forms of a single gene.
>
> A **dominant** allele will lead to a dominant characteristic when one or two dominant alleles are present.
>
> A **recessive** allele will lead to a recessive characteristic only when no dominant allele is present. This usually means that it is present on both chromosomes in a pair.

It is now understood that many characteristics are controlled by a pair of alleles – equivalent genes – found on a pair of chromosomes. The flower colour in Mendel's pea plants was controlled by a single gene. This gene had two possible forms called **alleles**, known as the **dominant** allele and the **recessive** allele. In pea plants the dominant allele produces a protein that forms red flowers. The recessive allele cannot produce this protein. So if one or both of the genes in the pair is the dominant allele the plant will have red flowers. It is only when there is no dominant allele (i.e. both recessive alleles) that the flowers will be white because there is no protein produced to turn them red. You often use a capital letter (e.g. R) to represent the dominant allele and a lower case letter (e.g. r) to represent the recessive allele.

A **recessive** allele will lead to a recessive characteristic only when no dominant allele is present. This usually means that it is present on both chromosomes in a pair.

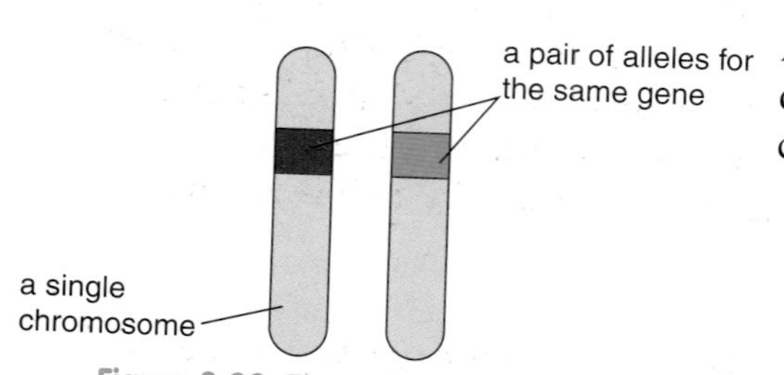

Figure 8.20 The red and the green section on the chromosomes represents a pair of alleles for the same gene.

24 Even though Mendel did not know about the nature of genes he helped to develop our current understanding of inheritance. Explain how he did this.

25 a) Use both types of genetic diagrams shown in Figure 8.21 to show the possible offspring when two Rr pea plants are crossed together.
b) Use your diagrams to explain how two red flowering pea plants can produce a seed that grows into a white flowering plant.

26 Humans may produce two types of earwax: a wet, sticky type or a dry, hard type. The wet type is caused by a dominant allele, E, and the dry type is caused by a recessive allele, e.
a) Write down the pairs of alleles that somebody with wet, sticky ear wax may have.
b) Explain why it is impossible to say what alleles a person who produces wet ear wax will have.
Rob and Jo are married. Rob has wet ear wax and Jo has dry ear wax. They have a child who produces wet ear wax.
c) What pair of alleles for ear wax will their child have?
d) Rob and Jo are planning on having another child. Draw a genetic diagram to show what possible pairs of alleles for ear wax their next child could inherit.

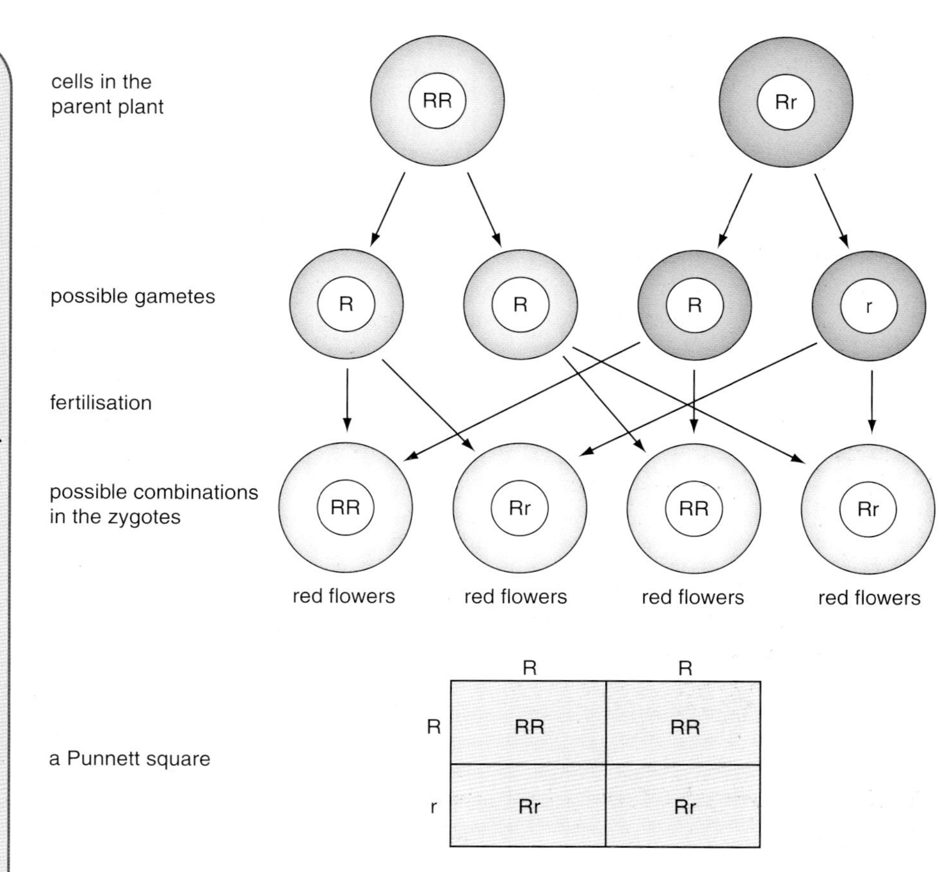

Figure 8.21 Two ways of drawing genetic diagrams to explain inheritance. Each shows all the possible combinations of alleles that could be passed on for a single gene.

How are some diseases inherited?

Some diseases are inherited because a defective gene from the parent is passed to their offspring. A defective gene is one that usually codes for a protein but no longer works because the order of bases has changed in one of the 46 chromosomes. This means that the instructions for making a specific protein no longer work. The lack of this protein causes the symptoms of the disease. **Huntington's disease** affects the nervous system. It causes gradual loss of co-ordination, jerky involuntary movements and often affects sufferers' sight and speech. People affected also develop dementia. Symptoms do not usually start to develop until people are in their 30s or 40s. Huntington's disease is caused by a dominant allele. A person will develop the disease if they inherit one dominant allele for this condition.

Cystic fibrosis is also inherited. It affects cell membranes and makes it difficult for a sufferer to absorb enough oxygen in the lungs. It is caused by a recessive allele. This means that a person needs to inherit two recessive alleles and no dominant alleles for this condition.

Huntington's disease is an inherited disease that affects the nervous system. It is caused by a dominant allele.

Cystic fibrosis is a disease that affects membranes of cells that line the lungs, gut and reproductive tract. It is caused by a recessive allele and therefore has symptomless carriers.

A **carrier** has one allele for an inherited disease caused by a recessive allele but does not have the disease. Two carriers can have children who inherit the disease.

27 A father had Huntington's disease and was Hh but his wife did not have the disease.
  a) What is the probability that their child would inherit the disease? Use a genetic diagram to work out your answer.
  b) If you were in this couple's situation, would you try to have children? Give reasons for your answer.

28 a) Name three people in the family tree in Figure 8.22 who were cystic fibrosis carriers.
  b) Name one person who suffered from the disease.
  c) Explain why Jess is unaffected by the disease.

29 Using Figure 8.22, draw a genetic diagram to show the possible outcomes of Claire having children with somebody who is a carrier of cystic fibrosis.

Huntington's disease

If you use an H to represent the dominant allele and an h for the recessive allele, the possible pairs of alleles are:

HH – this is a lethal combination. A fetus with this genotype would not develop.
Hh – this pair will also cause the disease.
hh – this pair will not cause the disease.

Cystic fibrosis

Cystic fibrosis is caused by a recessive allele. This means that a person can have the allele for this disease without actually suffering from the disease. In this case they would have one dominant and one recessive allele. A person with this pair of alleles is called a symptomless **carrier**. Figure 8.22 shows a family tree that has the cystic fibrosis allele in it. It is shown by cf.

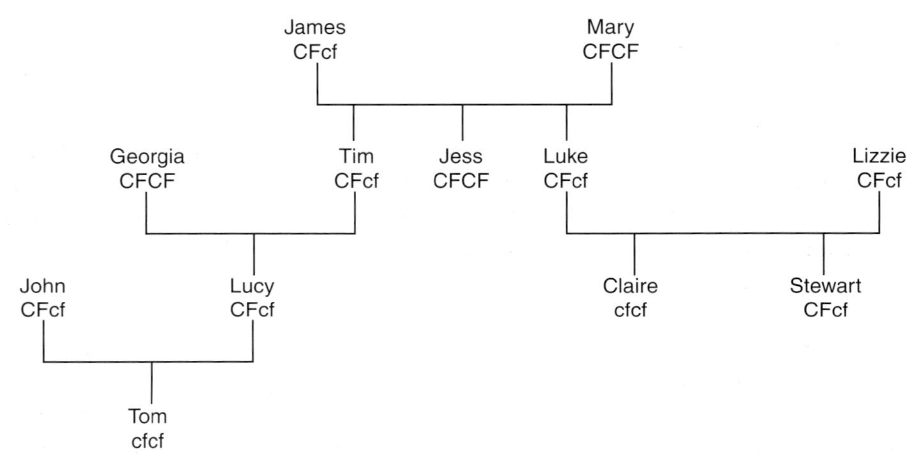

Figure 8.22 In this family the cystic fibrosis allele, cf, has passed through the generations.

Embryo screening

Embryo screening is a technique that can be used to identify some inherited diseases, such as Huntington's and cystic fibrosis. An embryo can be tested for these disorders and then only embryos that are free of the faulty allele are implanted into the woman's womb. You will remember from Chapter 4 that IVF (*in vitro* fertilisation) is often used when a couple cannot have children naturally. Embryo screening is only used as part of IVF and only offered to parents with a family history of inherited disorders.

Embryo screening is regulated by an organisation called the Human Fertilisation and Embryology Authority (HFEA). The HFEA receives many requests to extend the embryo screening technique to other disorders. Some genes are linked to various cancers. These genes do not cause these cancers directly. They only increase the chance of developing cancer, although not all people with the gene will get the disease. Embryo screening for genes linked to an increased cancer risk is permitted. However some people are concerned that in future the technique may be extended to allow parents to screen embryos for non-medical reasons.

30. It costs about £2000 to carry out IVF with embryo screening, which is twice as much as it costs without screening. With this in mind do you think that the use of embryo screening should be extended to include all births conceived by IVF? Give reasons for your answer.

31. Can you suggest why some people object to embryo screening, apart from financial reasons?

Summary

- ✓ The body controls its water and ion content to avoid cells becoming damaged.
- ✓ Body temperature is controlled by changes in the amount of sweat being produced by the **sweat glands** and in the amount of blood flowing through the arterioles to the **skin capillaries** (**vasodilatation** and **vasoconstriction**), and by shivering.
- ✓ Body temperature is monitored and controlled by the **thermoregulatory centre** in the brain.
- ✓ Blood glucose concentration is monitored and controlled by the **pancreas**. The pancreas produces **insulin** to move glucose from the blood into cells.
- ✓ **Diabetes** is a disease where the pancreas does not produce enough insulin. It is treated by controlling a person's diet and injecting with insulin.
- ✓ Urea and carbon dioxide are waste products of chemical reactions in the body.
- ✓ Carbon dioxide is removed from the body through the lungs and urea is removed by the kidneys.
- ✓ **Chromosomes** are molecules of **DNA** and are normally found in pairs.
- ✓ **Genes** are lengths of DNA responsible for making a specific protein.
- ✓ Body cells divide by **mitosis** to produce genetically identical cells for growth and for replacement of older cells.
- ✓ Cells in the reproductive organs divide by **meiosis** to produce genetically non-identical sex cells (**gametes**) with a single set of chromosomes.
- ✓ Gametes join at fertilisation to form a single body cell with a full set of chromosomes that divides by mitosis to form a new individual.
- ✓ Stem cells are a type of animal cell that remain **undifferentiated**. They can divide to produce a range of different types of specialised (**differentiated**) cells.
- ✓ Asexual reproduction by mitosis produces genetically identical offspring.
- ✓ Sexual reproduction leads to genetic variation because an individual inherits one allele in each pair from each parent.
- ✓ A person's sex is controlled by the **sex chromosomes**: XX is female and XY is male.
- ✓ **Alleles** are alternative forms of a gene for a specific characteristic.
- ✓ Some characteristics are controlled by a single pair of alleles which may be **dominant** or **recessive**. The effects of dominant and recessive genes can be demonstrated by **genetic diagrams**.

✓ Some disorders, such as **Huntington's disease** and **cystic fibrosis**, are inherited. Huntington's disease is controlled by a dominant gene. Cystic fibrosis is controlled by a recessive gene so there can be **carriers**. **Embryo screening** can be used to test for the faulty allele.

✓ There are ethical issues surrounding the use of stem cells from embryos, and in the use of embryo screening in IVF treatment.

EXAMQUESTIONS

1 a) Which part of the body monitors and controls body temperature? *(1 mark)*

b) Describe, in as much detail as you can, how the body responds to an increase in body temperature. *(2 marks)*

c) On a cold day people often appear to have very pale skin. Explain the scientific reason behind this observation. *(2 marks)*

2 a) Which organ in the body does not function properly in somebody who suffers from diabetes? *(1 mark)*

b) Explain how diabetes affects the body. *(3 marks)*

c) Describe two ways in which diabetes may be treated. *(2 marks)*

3 When a body builder trains, his muscles get bigger because more muscle cells are produced.

a) What type of cell division produces the new muscle cells? *(1 mark)*

b) What type of cell division produces his sperm cells? *(1 mark)*

c) Explain, in as much detail as you can, why the body needs two different types of cell division. *(4 marks)*

4 Fur length in rabbits is controlled by a single gene which has two alleles, L and l. A pair of rabbits bred together. The mother had long fur and the father had short fur. All their babies had long fur.

a) Give the alleles for fur length present in the body cells of:
i) the mother; ii) the father. *(1 mark)*

b) One of the offspring bred with a rabbit with short fur. Use a genetic diagram to show how some of their babies had long fur and some had short fur. *(3 marks)*

5 Medical researchers hope that human stem cells could be transplanted into patients to make new cells in patients with diseased or damaged body tissue. This could lead to treatments for diseases such as Alzheimer's and Parkinson's, or for spinal cord injuries.

Most human stem cells used in research come from embryos created but not used at *in-vitro* fertilisation clinics. The cells are taken from embryos a few days old and the embryo is destroyed. The cells are then grown in the lab. At the moment no-one is quite sure how to make the stem cells form particular types of tissue – such as nerve rather than brain.

Because the cells come from embryos, the research is controversial. To some people, an embryo is a human and has a right to life. Those who support the use of stem cells from embryos say that a four-day-old embryo has not yet developed to a stage where human life has begun.

Stem cells are also found in adult bone marrow and in a baby's umbilical cord blood. Some people say that embryonic stem cells are not the only hope of a treatment for, say, Parkinson's disease. Unfortunately, those stem cells may be limited in the cell types they can develop into. Embryonic stem cells can grow into 200 different types of specialised cell.

a) Explain why researchers think stem cells collected from embryos could lead to medical treatments. *(1 mark)*

b) Use the information in the passage to suggest why people may be concerned about the source of stem cells used in medical research. *(1 mark)*

c) A patient who is paralysed says that whatever the ethical concerns, they think the research should continue so that a cure can be found. Explain the difficulties researchers have to overcome for this therapy to become available to patients in the future. *(2 marks)*

Chapter 9
Why are diffusion and active transport so important to living things?

At the end of this chapter you should:

- ✓ be able to explain how dissolved substances move by diffusion and active transport;
- ✓ understand that active transport requires energy whereas diffusion does not;
- ✓ be able to describe how the lungs are specialised to absorb oxygen and dispose of carbon dioxide;
- ✓ know that the lungs are protected by the ribcage in the thorax and separated from the abdomen by the diaphragm;
- ✓ be able to describe how the small intestines are specialised to absorb nutrients;
- ✓ know how carbon dioxide enters the leaves of plants by diffusion;
- ✓ know how water and minerals are absorbed through the roots of plants;
- ✓ be able to explain how water leaves a plant by transpiration and how this can be controlled;
- ✓ be able to describe how the kidneys remove toxic substances from the blood;
- ✓ be able to explain how the kidneys maintain the correct concentrations of water, sugar and dissolved ions in the blood;
- ✓ understand how kidney failure can be treated either by a dialysis machine or with a transplant;
- ✓ be able to evaluate the advantages and disadvantages of treating kidney failure by dialysis or kidney transplant.

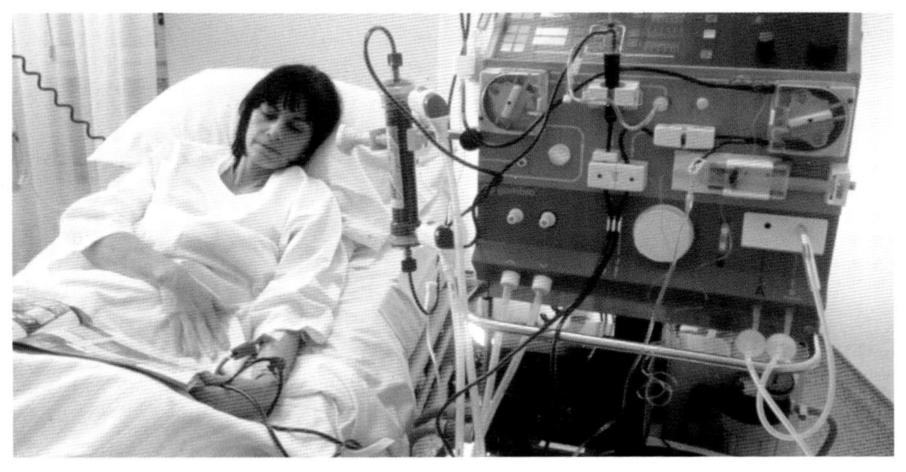

Figure 9.1 In all these photos, substances are getting into and out of living things. Once inside a living plant or animal, substances have to be moved around. This can take place by diffusion and active transport.

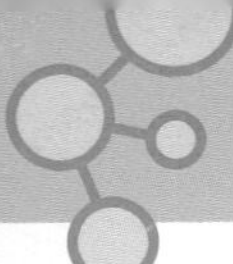

What is the difference between diffusion and active transport?

From Chapter 5, you should know how gases and dissolved substances can move by diffusion. The particles in a gas or a liquid are constantly moving. If there is a high concentration of particles in one part of a gas or solution, they will tend to spread out and move to where there is a lower concentration of these particles. This movement of particles from a region of higher concentration to one of a lower concentration is called diffusion. In humans, oxygen diffuses into a cell from the blood through the cell membrane so that the cell can respire (see Section 5.2). Oxygen is at a relatively high concentration in the blood. It moves by diffusion to where it is at a lower concentration inside the cell. In moving from a place of higher concentration to one of lower concentration, we say the oxygen is moving down a **concentration gradient**.

Diffusion is the movement of particles in a gas or solution down a concentration gradient. A **concentration gradient** exists between two areas when one area has a higher concentration of a substance than the other area.

1 Visking tubing is a synthetic material with tiny holes in it. These holes are big enough for small molecules, such as water, to pass through.

a) If a length of visking tubing is filled with pure water and then placed in a beaker of sugar solution, what will happen?

b) Explain your answer to part **a)**.

Figure 9.2 Oxygen molecules move down a concentration gradient, from a higher concentration in the blood to a lower concentration in a cell.

When diffusion occurs in living things, substances move from regions of higher concentration to regions of lower concentration. However, living organisms sometimes need to move a substance from an area where its concentration is very low to an area where there is already a higher concentration of the substance. Consider a plant taking minerals into its roots from the soil. Some minerals, such as phosphate ions, have a very low concentration in the soil and a higher concentration inside the root cells. However, the plant still needs to take in phosphate ions from the soil. In this situation, diffusion is useless because the phosphate ions would move out of the root cells into the soil, down a concentration gradient.

Instead, the plant uses a process called **active transport** to move phosphate ions from the soil into the root cell. Active transport is used in living organisms to move substances through a membrane up a concentration gradient, from a lower to a higher concentration. Unlike diffusion, active transport cannot rely on the random movement of particles to move them through a membrane. Instead, chemicals in the membrane bond with particles on the outside of the cell, move them across the membrane and then release them into the cell. This process uses energy that the cell releases during respiration. Humans use active

Active transport is the movement of particles through a membrane, up a concentration gradient. The energy required for active transport is obtained from respiration.

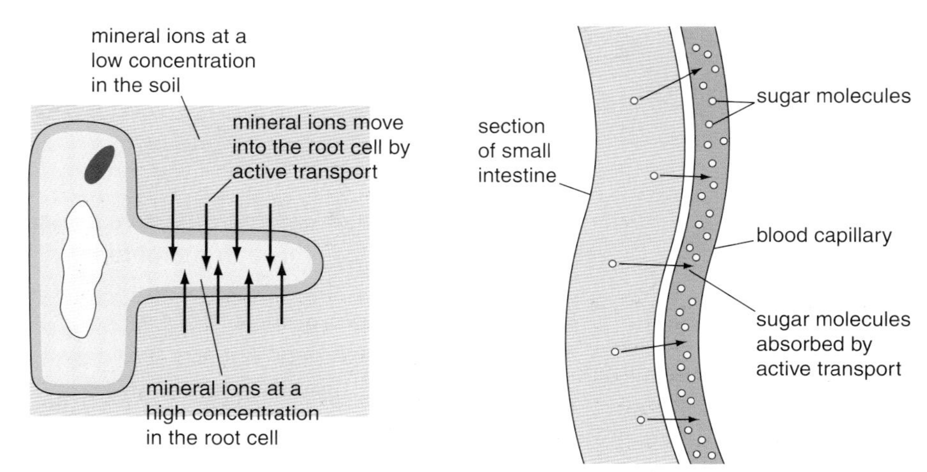

Figure 9.3 Plants and animals use active transport to move substances from regions of low concentration to regions of higher concentration up concentration gradients. Plants move mineral ions into their roots using active transport, and animals sometimes use this process to move sugars from the intestines into the bloodstream.

transport to absorb sugar from the small intestines where they have a very low concentration of sugar into the blood where there is a higher concentration.

One way to understand the difference between diffusion and active transport is to compare them to riding a bike down and up a hill. If you are on a bike at the top of a hill, you can roll down the hill without using any of your own energy. This is like diffusion. You use no energy and move down the gradient (slope). In contrast, active transport is like riding a bike up the hill. Here you are cycling up a gradient and have to use some of your own energy. You obtain this energy from respiration, just like a cell using active transport.

Figure 9.4 Diffusion is like cycling downhill: substances move down a concentration gradient, using no energy. Active transport is like cycling uphill: substances move up a concentration gradient, using up energy in the process.

2. Why do substances not move through visking tubing by active transport?
3. When plants photosynthesise they produce sugar inside cells, and from here the sugar is transported around the plant. The sugar is often at a lower concentration inside the cells than in the tubes that transport it around the plant. However, plants can move the sugar from their cells into the transport tubes.
   a) Describe the concentration gradient that exists between the plant cells and transport tubes.
   b) How do you think the sugar is moved from the cell into the tube so that it can be transported around the plant?
4. Sugar can be absorbed very efficiently into our blood system from the small intestine. After a meal, there may be a high concentration of sugar in the small intestine. But even when the concentration of sugar is lower in the small intestine than in the blood, it is still absorbed into the bloodstream.

   Explain why membranes in the small intestine use both diffusion and active transport to absorb sugar.

9.2 How are our breathing and digestive systems specialised for absorbing substances?

Our breathing and digestive systems both have the ability to transfer substances through membranes. The breathing system takes air into the lungs so that oxygen can be absorbed into the bloodstream. At the same time, carbon dioxide passes out of the bloodstream into the lungs before being exhaled. The digestive system absorbs nutrients from our food into the bloodstream. Both of these systems have specialised features that increase their ability to transfer materials through their surface membranes.

The breathing system

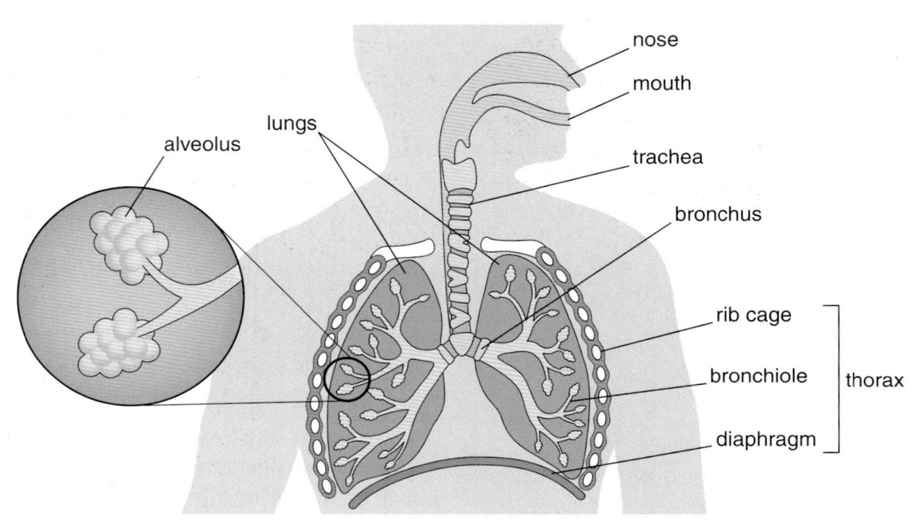

Figure 9.5 The lungs are in the upper part of the body (thorax) and are protected by the ribcage. The diaphragm separates the thorax from the abdomen.

Your breathing system moves gases into and out of your body. When you breathe in, your lungs inflate with air. When you breathe out, most of the air passes out of your lungs. While each breath of air is in your lungs, your body must absorb as much oxygen as possible into the blood. At the same time, as much carbon dioxide as possible must be transferred from your bloodstream into your lungs. The carbon dioxide is then lost from your body into the air when you breathe out. This movement of oxygen from the air into your bloodstream and carbon dioxide in the opposite direction is called **gas exchange**.

Gas exchange takes place in the alveoli. This involves oxygen moving from the air into the blood and carbon dioxide moving from the blood into the air.

There is always a higher concentration of oxygen in the air than in the blood. So, oxygen can move by diffusion down a concentration gradient from the air in the lungs into the blood. Carbon dioxide is produced by respiration in body cells and transported to the lungs in the blood. The concentration of carbon dioxide in the blood is greater than that in the air so it diffuses down a concentration gradient from the blood to the air in the lungs.

Alveoli are tiny pockets in the lungs that provide a very large, moist surface with a rich supply of blood capillaries so that gases can diffuse readily into and out of the blood.

Notice in Figure 9.5 that the lungs are made up from lots of **alveoli** (singular alveolus). In fact, the average adult has about 600 million alveoli. They are tiny air sacs that give the lungs a spongy texture. Alveoli make the lungs very effective at gas exchange because they have a number of specialised features (Figure 9.6).

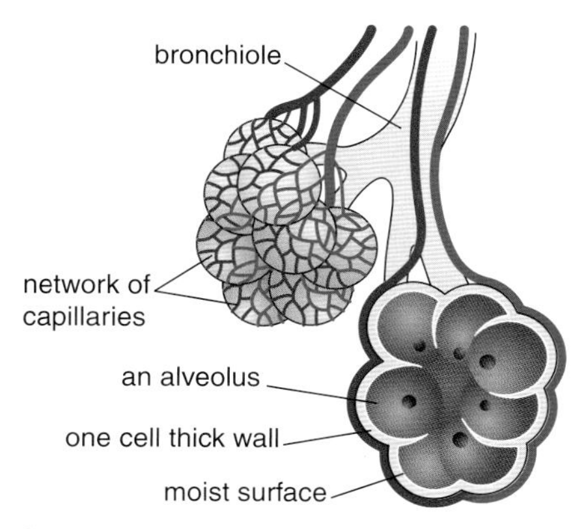

Figure 9.6 A cluster of alveoli showing the specialised features for gas exchange.

So, what are the features of alveoli that increase gas exchange?

- **Large surface area**. If you could flatten out all the alveoli in your lungs they would cover about half a tennis court. The large surface area makes it easier for the gases to come into contact with the membrane surface and then pass across it.
- **Moist surface**. Oxygen and carbon dioxide must dissolve in liquid before they can diffuse through the alveolus wall. Cells in the alveolus wall produce a liquid in which the gases can dissolve, allowing diffusion to take place.
- **Thin walls**. The walls of an alveolus are only one cell thick. This means that the gases do not have to diffuse far for exchange to occur.
- **Rich supply of blood capillaries**. Capillaries are wrapped around the walls of the alveoli so that gas exchange between blood and air can take place easily. Carbon dioxide diffuses out of the blood and oxygen diffuses in, while the blood continuously moves through the capillaries and around the body. This continuous flow helps to maintain a sufficient concentration gradient for oxygen and carbon dioxide to diffuse between the air and the blood.

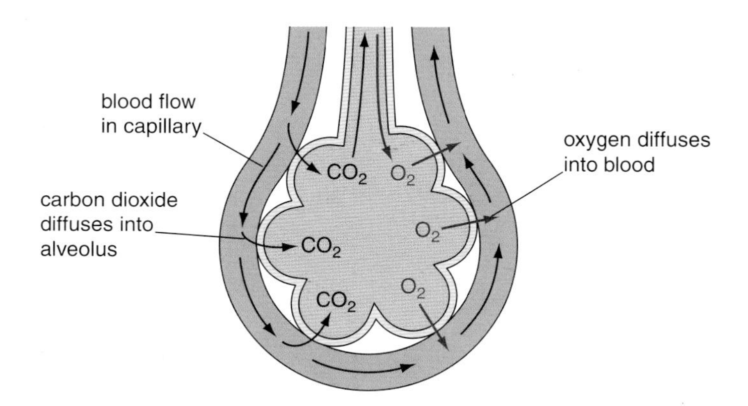

Figure 9.7 Gas exchange of oxygen and carbon dioxide in an alveolus. Oxygen diffuses from a higher concentration in the alveolus to a lower concentration in the blood. Carbon dioxide diffuses from the blood into the alveolus.

5. The lungs are organs that are vital for life. How are they protected?
6. At high altitudes the air pressure is lower. This means that mountaineers who climb to high altitudes take in less oxygen with each breath than they would at sea level. How will this affect the rate at which oxygen is absorbed in their lungs? Explain your answer.

The digestive system

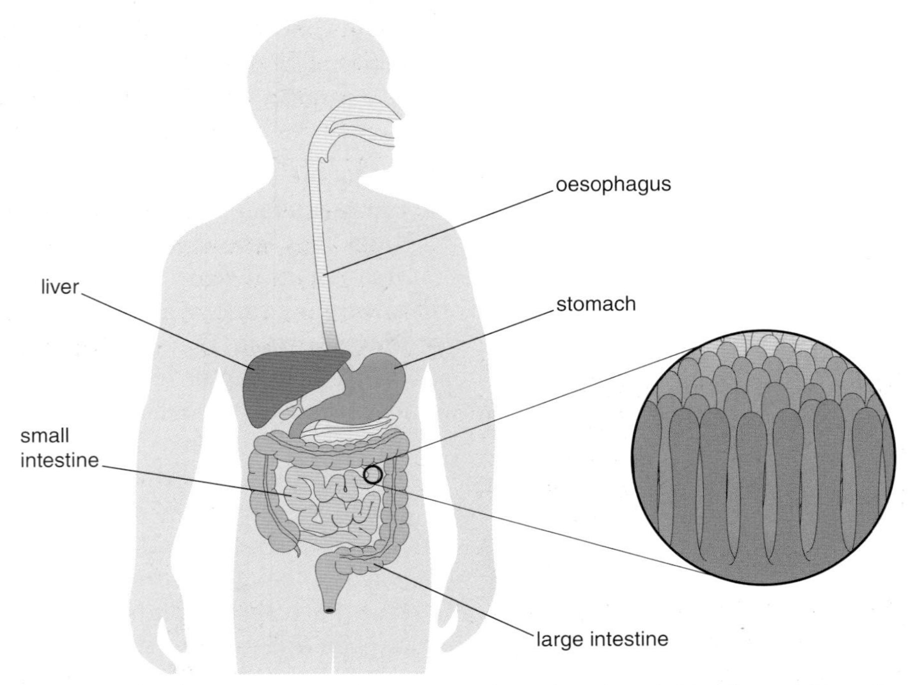

Figure 9.8 The digestive system showing an enlarged section of villi in the small intestine

Villi in the small intestine provide a large surface area with an extensive network of capillaries so that the products of digestion can be absorbed into the bloodstream by diffusion and active transport.

In your digestive system large nutrient molecules such as proteins and starch are digested to produce smaller molecules such as amino acids and sugar. These smaller molecules are then absorbed into the bloodstream. Absorption takes place through tiny structures called **villi** in the small intestine. The process involves moving dissolved substances through cell membranes in the villi. Villi in the small intestine work in a similar way to alveoli in the lungs. They provide a very large surface area and have an extensive supply of capillaries enabling the small intestine to absorb nutrients very effectively.

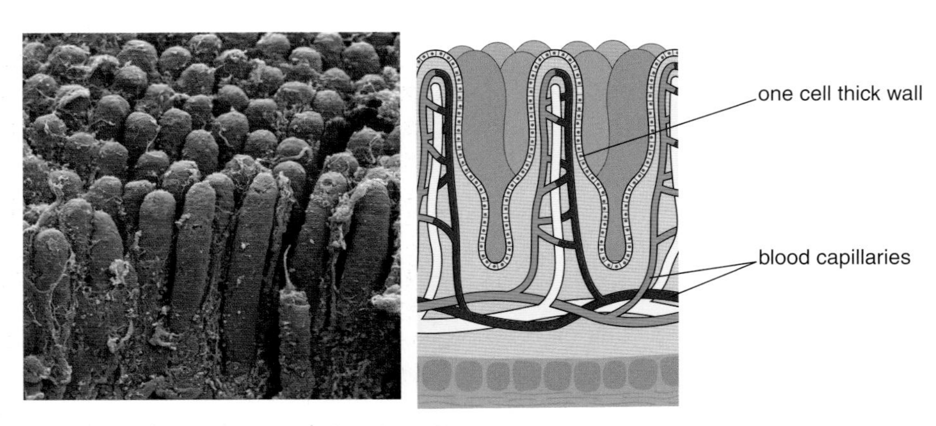

Figure 9.9 Villi line the inside surface of the small intestine. Their thin surface membranes and good blood supply enable nutrients to pass easily from the small intestine into the blood.

The alveoli in the lungs use only diffusion to move oxygen and carbon dioxide. Villi need to use both diffusion and active transport to absorb nutrients. After a meal the concentration of nutrients in your small intestine will soon increase. At this stage their concentration around the villi will be greater than their concentration in the blood inside the villi. This means that the nutrients can move down a concentration gradient into the blood by diffusion.

When the concentration of nutrients in the small intestine gets very low, they have to move up a concentration gradient to get into the blood. This means that the cells in the villi wall must use active transport to absorb the remaining nutrients. The result is that most of the nutrients are absorbed into the bloodstream but some energy must be expended in the process.

7. A large surface area is essential for effective absorption. Explain how this is achieved in the breathing system and the digestive system.
8. Why do the lungs only need to use diffusion to maintain gas exchange?
9. Why is it important that the small intestine uses both diffusion and active transport to absorb nutrients?

small intestine
capillary
nutrients absorbed by diffusion
a)
nutrients absorbed by active transport
b)

Figure 9.10 The small intestine is very effective in absorbing nutrients because it uses both diffusion and active transport. Diagram a) shows particles being absorbed by diffusion. When their concentration drops below their concentration in the blood, the particles are absorbed by active transport. This is shown in diagram b).

How do plants absorb and transport substances?

Like animals, plants need to take in, transport and get rid of substances to stay alive. The roots and leaves of plants are adapted to carry out these exchanges of substances effectively. Three exchange processes are particularly important:

- The carbon dioxide needed for photosynthesis enters a plant through its leaves (Figure 9.11).
- The oxygen produced during photosynthesis is lost from the leaves along with some water.
- Plants absorb water and mineral ions from the soil through their roots (Figure 9.12).

If you magnified the inside of a leaf you would see a number of different structures. Each of these carries out a specific function. These structures make the leaf very efficient at transporting substances.

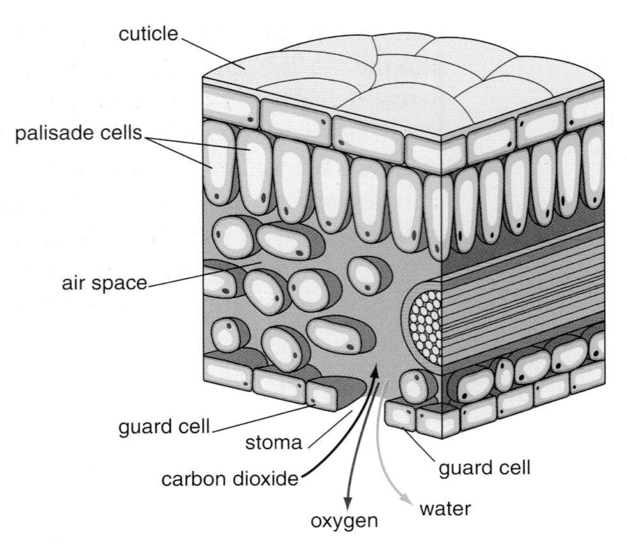

Figure 9.11 A cross section of a leaf showing the movement of materials through the stomata

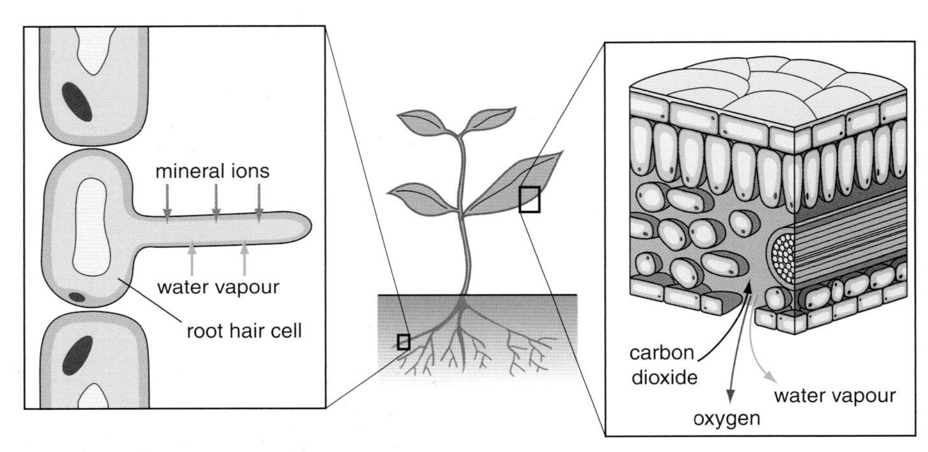

Figure 9.12 Root hair cells give the roots a large surface area for absorption and internal air spaces give leaves a large surface area for gas exchange.

Figure 9.11 shows how carbon dioxide gets into a leaf through holes called stomata and then moves through the air spaces inside the leaf to reach the cells (see Chapter 6). Photosynthesis takes place mainly in the palisade cells. Palisade cells contain many chloroplasts. Photosynthesis takes place in the chlorophyll inside the chloroplasts. This process needs a good supply of carbon dioxide and water to keep going. Leaves are usually flat with a large surface area. A large surface area means lots of stomata so that carbon dioxide can be absorbed effectively. Oxygen is produced during photosynthesis, and this gas diffuses out of the plant through the stomata into the air.

Plants also lose water vapour through their stomata (singular stoma). There is usually a higher concentration of water inside the leaf than

Transpiration is the loss of water from a plant, mainly through stomata in the leaves. Transpiration is more rapid in hot, dry and windy conditions.

outside, so water not required for photosynthesis diffuses out of the leaf into the air. This is called **transpiration**. Loss of water by transpiration is fine if there is a good supply of water into the roots by osmosis, which is a kind of diffusion (see page 82). However, if a plant loses water faster than it is replaced through the roots, the plant will wilt and its cells may be damaged. Fortunately, the size of the stomata can be controlled by specialised cells called guard cells that enclose them (Figure 9.13). If a plant is losing water too quickly, the guard cells change shape so that the stomata become smaller, and the rate of transpiration is reduced.

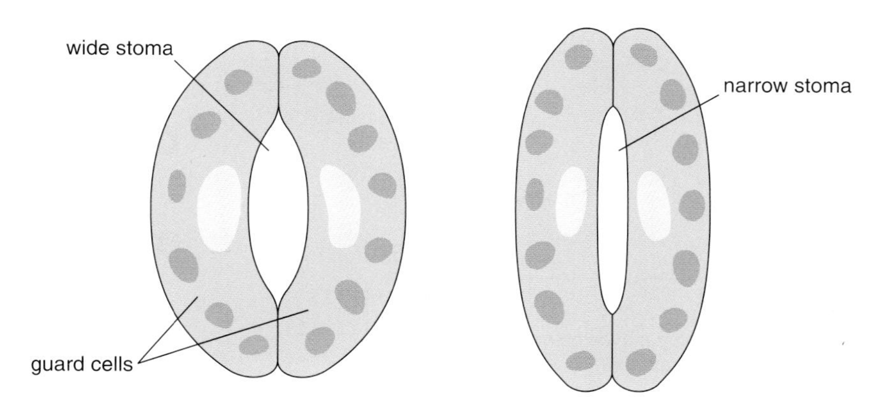

Figure 9.13 Stomata become smaller if a plant needs to reduce the amount of water being lost by transpiration.

10. Look at Figure 9.11. Gas exchange takes place through the stomata in leaves.
    a) Which gases are exchanged?
    b) What process is used to move these gases into and out of the leaf? Explain your answer.
11. Look at Figure 9.12. Water and mineral ions are being absorbed into a root hair cell. Water is moving from an area of high concentration in the soil to a lower one inside the root hair cell. On the other hand, minerals must often move from an area of lower concentration in the soil to one of higher concentration inside the cell.
    a) How do root hair cells increase the rate at which a plant can absorb water and mineral ions?
    b) What process is used to absorb water into root hair cells?
12. How are mineral ions absorbed when they are at a lower concentration in the soil than in the root hair cell?

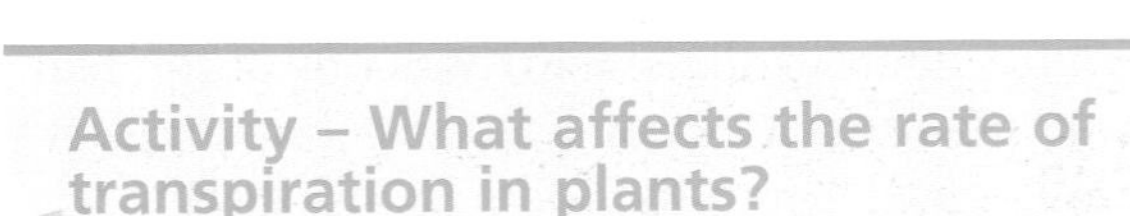

Activity – What affects the rate of transpiration in plants?

The rate of transpiration in plants can be measured using a potometer. The plant draws up the water it needs through a very narrow tube. An air bubble in the tube moves as the water is absorbed. By measuring how far the air bubble moves each minute you can compare the transpiration rate under different conditions.

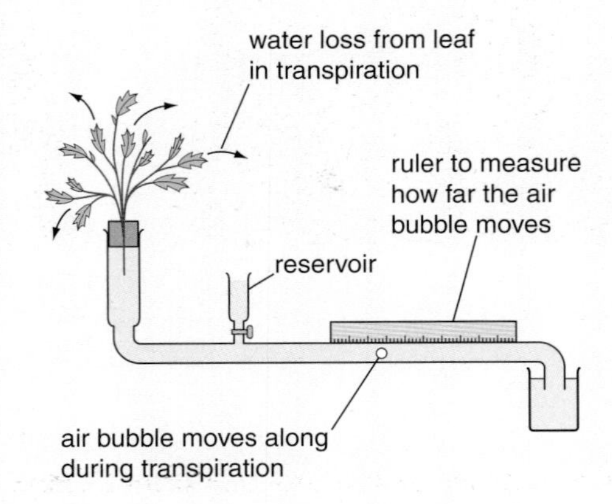

Figure 9.14 Measuring the rate of transpiration with a potometer

Zoe carried out an investigation using a potometer to find out how different conditions affected the rate of transpiration. She carried out three experiments under the following conditions:

1 on a bench in the laboratory;
2 in the laboratory with a fan blowing over the plant;
3 in a warm incubator.

Her results are shown in Table 9.1.

❶ What should Zoe do to check that her results are reliable?
❷ Plot a graph of Zoe's results. Choose the scales so that all the results can be shown on one graph.
❸ What conclusions can you draw about the effect of different conditions on the rate of transpiration?
❹ Sketch on your graph the line you would expect to get if you carried out the experiment using a hair drier set to warm instead of a fan.
❺ One of the results appears to be anomalous (i.e. it doesn't fit with the others). Which one is it? Suggest one reason for the anomalous result.
❻ Design an experiment to investigate the effect of humidity on the rate of transpiration. (Humidity means the amount of water vapour in the air.)
❼ What effect do you think increasing the humidity will have on the rate of transpiration?

	Distance the bubble moved after each minute in mm									
Time in min	**1**	**2**	**3**	**4**	**5**	**6**	**7**	**8**	**9**	**10**
Experiment 1	2	3	5	7	10	11	13	15	18	20
Experiment 2	13	24	38	49	73	75	89	104	118	132
Experiment 3	5	12	18	24	30	35	41	47	53	58

Table 9.1

9.4 How do the kidneys help to maintain internal conditions in the body?

Your kidneys remove toxic substances from your body. These substances are excreted in your urine. To do this, the kidneys use diffusion, osmosis and active transport. If the kidneys are damaged, some of the toxic substances may not be removed and this can lead to serious illness and even death. Fortunately, kidney failure can now be treated. But, before looking at the treatments for kidney failure, we need to understand how a normal kidney works.

⓭ What would happen to a person's blood if kidney failure went untreated?

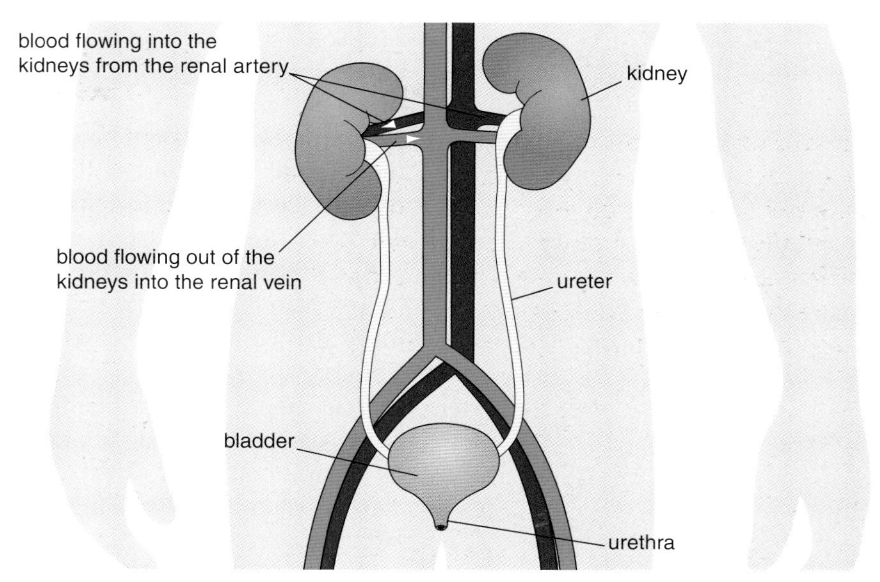

Figure 9.15 The kidneys are attached to the circulation system by arteries and veins. They produce urine that is transported to the bladder via the ureter and then out of the body by the urethra.

Nephrons are the structures in the kidney that filter the blood and release waste (urea, excess ions and water) as urine.

The kidneys use a sequence of processes to produce urine. These processes take place in the **nephrons**. There are about one million nephrons in each kidney.

In the nephrons, the process of removing waste and making urine takes place in several stages (Figure 9.16).

- The blood first passes into the capillaries in the glomerulus under high pressure. At this stage, the blood contains red and white blood cells, proteins, platelets, water, urea, dissolved sugar and salts. Urea is the main toxic substance that has to be removed from the body (Section 8.2).

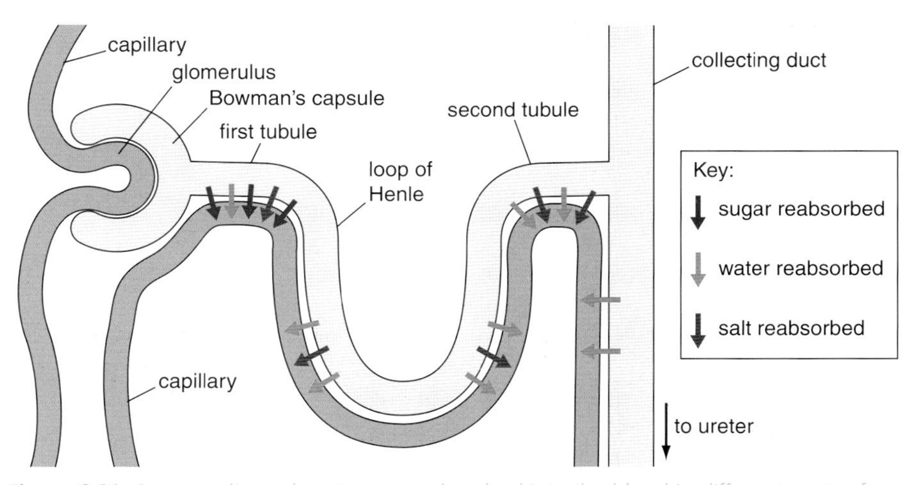

Figure 9.16 Sugar, salts and water are reabsorbed into the blood in different parts of a nephron.

- The walls of these capillaries have tiny holes that act as filters. They allow the water and dissolved salts, sugar and urea to pass out of the blood and into Bowman's capsule, a sac that surrounds the capillaries. This creates a problem because the body must get all of the sugar and most of the water and salts back into the blood. It only needs to excrete the urea, some salts and some water.
- This mixture in the sac then enters the first tubule, where a process called reabsorption begins. In this tubule, sugar is reabsorbed into the blood. Initially the process occurs by diffusion, but active transport must also be used to reabsorb all the sugar.
- The reabsorption process continues in the loop of Henle and the second tubule until the required amounts of water and dissolved salts have been reabsorbed. Water is reabsorbed by osmosis and the salts by a combination of diffusion and active transport.
- This leaves water, small amounts of salt and urea in the collecting duct. Here, adjustments are made to the amount of water that is finally reabsorbed into the blood. If the body is dehydrated, the collecting duct responds to the hormone ADH (anti-diuretic hormone) by increasing the amount of water that passes through its wall and back into the blood (see Section 1.2). If the body is well hydrated, then a drop in ADH causes the collecting duct to reduce the amount of water that is reabsorbed, so that more water can pass out of the body in the urine.
- Finally, a mixture of urea, excess salts and water passes along the ureter and into the bladder. This mixture is urine, which is excreted when you go to the toilet.

14 Why do you think that blood cells, platelets and proteins are not filtered out of the blood in the glomerulus?

15 Why does some sugar need to be reabsorbed by active transport?

16 Explain how the collecting duct plays a crucial role in controlling the amount of water in the blood.

17 a) State the differences between blood entering and blood leaving the kidneys.
b) Explain each of the differences.

These processes in our kidneys are complicated, but it is crucial that they take place continuously. This ensures that urea is removed from the body and that water is maintained at the right level in the blood. Both of these functions are disrupted if the kidneys are damaged by injury or disease.

Part of the nephron	Process
Glomerulus	Water, urea, sugar and salts are filtered out of the blood.
First tubule	Sugar is reabsorbed into the blood by diffusion and active transport. Some water is reabsorbed into the blood by osmosis. Salts are reabsorbed into the blood by active transport.
Loop of Henle	More water and some salts are reabsorbed into the blood.
Second tubule	More water and some salts are reabsorbed into the blood.
Collecting duct	Water is reabsorbed into the blood. More water is reabsorbed if the body is dehydrated. The remaining mixture of water, urea and some salts forms urine and passes into the ureter.

Table 9.2 A summary of the processes that take place in the kidney

Treating kidney failure

People may suffer from kidney failure for various reasons. Kidney failure can result from injury, infectious diseases or conditions that have been inherited. Kidney failure is treated in one of two ways: either by using a kidney dialysis machine or by a kidney transplant. There are a number of things to consider before a decision can be made about the type of treatment a patient will receive.

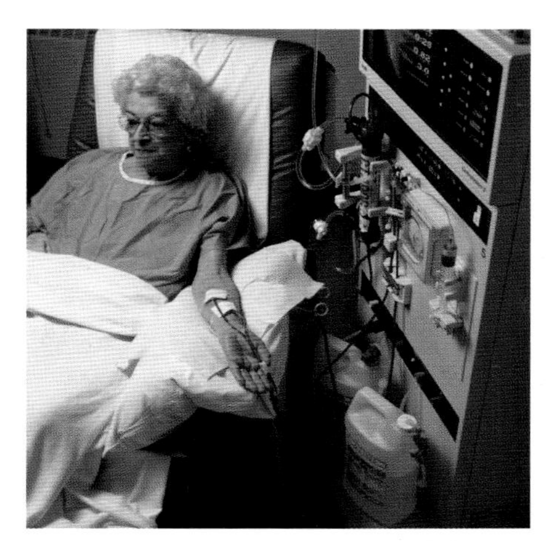

Figure 9.17 A patient with kidney failure using a kidney dialysis machine

Kidney dialysis

Dialysis is used when the kidneys' function is at a level that would result in death if the patient were left untreated. Someone using dialysis will normally use the machine three times a week for about four hours each time. Blood is taken from a vein in the patient's arm and fed into the machine. The blood flows alongside a partially permeable membrane. On the other side of the membrane is a liquid called dialysis fluid. This contains glucose and salts at the same concentration as the blood but no urea. This means that useful substances (glucose and salts) are not lost from the blood, but harmful urea diffuses from the blood into the dialysis fluid across the partially permeable membrane. The treated blood is then returned to a blood vessel in the patient's arm.

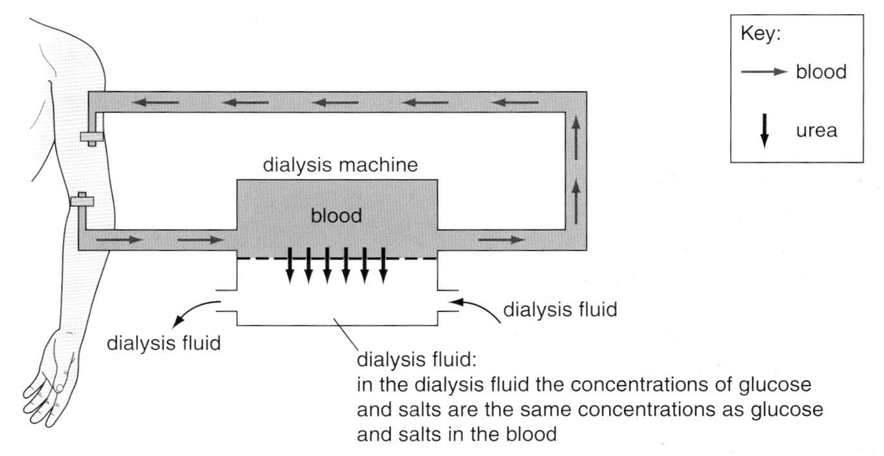

Figure 9.18 A dialysis machine removes urea from the blood without removing useful substances.

18 a) For how long would a dialysis patient normally be attached to a dialysis machine each week?
b) How do you think this will affect the patient's lifestyle?

19 Blood entering the dialysis machine has a high concentration of urea. Explain how the urea is removed from the blood.

20 It is essential that the machine does not remove useful substances such as glucose. Explain how this is achieved.

21 How does the way that a dialysis machine works differ from a real kidney?

Kidney transplant

A kidney transplant involves a major operation to replace a damaged kidney with a healthy one from a donor. A donor may be a living person, often a close relative, who gives one of their kidneys to the patient. This is possible because people can lead a perfectly healthy life with only one kidney. A donor may also be someone who has died after previously agreeing to donate body organs.

22. What are the advantages of a kidney transplant compared with treatment using a dialysis machine?
23. Apart from the medical problems, suggest one other problem that someone who has had a transplant may encounter.
24. When a kidney becomes available for transplant, there will always be more than one suitable recipient. In these cases, doctors must decide who will receive the organ. With a partner, brainstorm a list of factors that you think the doctors should take into account. Then put these factors in order of importance.

Have a look at www.uktransplant.org.uk/ukt/ and follow links to 'about transplants, organ allocation, kidney (renal)' to see how the NHS decides who will receive a donated organ.

The operation takes about two hours and, if successful, replaces the need for dialysis. However, there may be complications. The patient's body may reject the donated kidney and his / her immune system may start to treat the organ as an infection. The recipient and donor will have had a 'tissue test' to verify that the two are compatible before the operation. If the kidney starts to show signs of being rejected, the patient is treated with drugs that suppress the immune system. It is very rare that the donor and recipient are a perfect match, but a close match means that fewer immune-suppressing drugs will be needed. The biggest problem facing a patient who is suitable for a transplant is the lack of suitable donors. This means that the patient may have to wait many months before an operation is possible.

How much does treatment cost?

Both kidney dialysis and kidney transplants are expensive. It costs the National Health Service approximately:

- £21 000 to keep a patient on dialysis for just one year;
- £17 000 for a transplant operation;
- £5000 every year after a transplant to maintain the health of a patient.

25 How much does a transplant patient cost the NHS in the first year of treatment? Remember to include the cost of both the operation and drugs.

26 a) How much does it cost to treat a transplant patient for the five years from immediately after their operation?
b) Compare this with the cost of treating a patient by dialysis for five years.
c) How much money does the NHS save in 5 years if it treats kidney failure with a transplant operation?

27 About 20 000 patients are treated by dialysis in the UK. If 5000 donated kidneys could be found, how much money would this save the NHS over five years?

28 Currently about 20% of the UK population are registered as potential donors if they die. What do you think could be done to encourage more people to register as organ donors?

Summary

- ✓ Dissolved substances move down a **concentration gradient** by diffusion.
- ✓ Substances can be absorbed against a concentration gradient by **active transport**. This requires energy from respiration.
- ✓ The lungs have thousands of **alveoli** that provide a large surface area for absorbing oxygen into the blood.
- ✓ Alveoli have a large, moist surface area covered in capillaries so that gases can readily diffuse into and out of the blood (**gas exchange**).
- ✓ The small intestine has **villi** that provide a large surface area with an extensive blood supply to absorb the products of digestion.
- ✓ Absorption in the small intestine takes place by diffusion and active transport.
- ✓ Carbon dioxide enters plants through stomata in the leaves by diffusion.
- ✓ Water and mineral ions enter a plant through the root hair cells.
- ✓ The movement of water out of a plant's leaves is called **transpiration** and is more rapid in hot, dry and windy conditions.
- ✓ A plant controls the rate of transpiration by altering the size of its stomata.
- ✓ Kidneys remove urea from the blood and maintain the correct concentrations of sugar, water and salts in the blood. **Nephrons** are structures in the kidney that filter the blood and release waste (urea, excess ions and water) as urine.
- ✓ People who suffer from kidney failure can be treated by using a dialysis machine or by having a kidney transplant operation.
- ✓ Dialysis removes urea from the blood and restores the concentration of other dissolved substances to normal levels.
- ✓ A transplanted kidney may be rejected by the patient's immune system. Tissue-type matching and treatment with immune-suppressing drugs help to prevent this.

EXAMQUESTIONS

1. Oxygen is absorbed from the air into the blood in the lungs.
   a) What is the name of the process through which oxygen is absorbed into the blood? *(1 mark)*
   b) Describe two features of the lungs that make them very effective at absorbing oxygen. *(2 marks)*
   c) Explain how the arrangement of blood vessels in the lungs ensures that oxygen is constantly absorbed into the blood. *(2 marks)*

2. a) Water enters a plant at the roots and exits a plant through the leaves. With reference to specific parts of the plant, explain how this process takes place. *(5 marks)*
   b) If plants are losing too much water, they can reduce the rate at which water leaves the plant. Explain how this reduction in water loss takes place. *(3 marks)*
   c) Name two conditions that increase the rate of water lost by a plant. *(2 marks)*

3. A person who suffers from kidney failure may be treated with a dialysis machine or with a kidney transplant.
   a) Explain how a kidney dialysis machine works. *(4 marks)*
   b) Give one advantage and one disadvantage of dialysis treatment. *(2 marks)*
   c) Give two advantages of a kidney transplant as the treatment for kidney failure. *(2 marks)*

4. Healthy kidneys remove urea from the blood, but maintain the correct levels of sugar and salt. Explain how this takes place. *(6 marks)*

5. 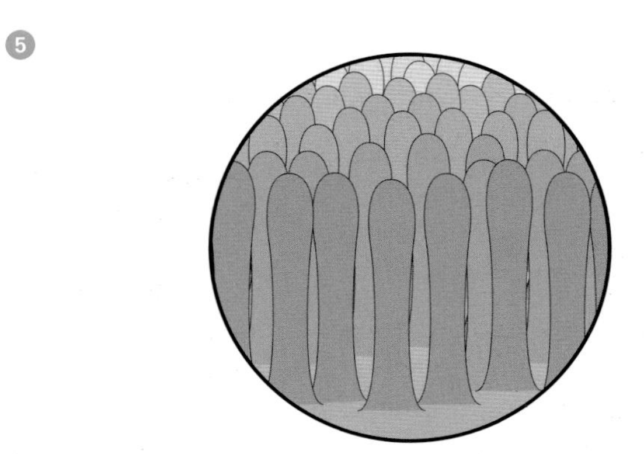

   Figure 9.19 Villi in the small intestine

   a) What is the function of villi in the small intestine? *(1 mark)*
   b) The concentration of glucose in the blood surrounding the small intestine is often much higher than the concentration of glucose in the small intestine itself. However, glucose is still absorbed into the blood under these conditions. Explain how this takes place. *(2 marks)*
   c) People who suffer from a disease called coeliac have villi that are much shorter than normal. They often suffer from weight loss and a lack of energy.
   Explain why coeliac sufferers have these symptoms. *(3 marks)*

Chapter 10
How are materials transported around the body and how is this affected by exercise?

At the end of this chapter you should:

- ✓ be able to describe how blood flows around the body;
- ✓ understand the role of arteries, veins and capillaries in the circulation system;
- ✓ be able to explain how blood transports carbon dioxide, urea and the soluble products of digestion;
- ✓ know how red blood cells are adapted to transport oxygen very efficiently;
- ✓ understand how energy is released from glucose in muscle cells by aerobic and anaerobic respiration;
- ✓ know that muscles use the energy released during respiration to contract;
- ✓ be able to describe how exercise changes the heart rate, breathing and arteries;
- ✓ know how these changes affect the supply and removal of substances to and from the muscles;
- ✓ know how muscles use glycogen during exercise;
- ✓ be able to interpret data about the effects of exercise on the body;
- ✓ be able to explain how muscles respond to long periods of vigorous exercise;
- ✓ understand how an oxygen debt builds up in muscles during anaerobic respiration owing to the formation of lactic acid;
- ✓ be able to explain how the body 'repays' an oxygen debt.

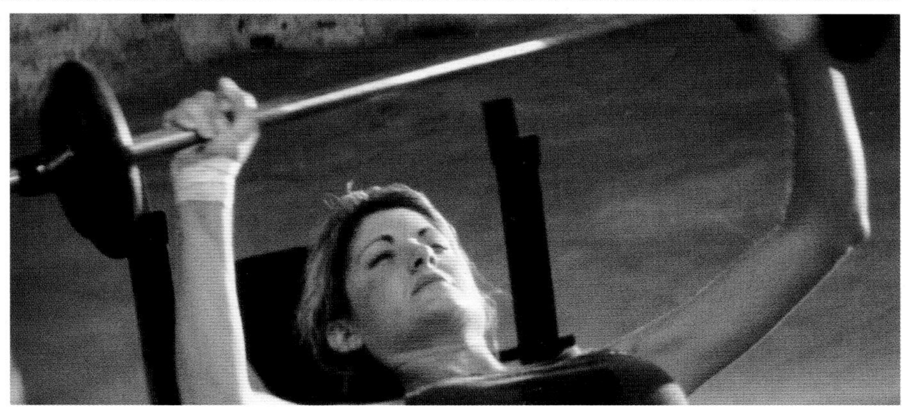

Figure 10.1 Exercise is a crucial part of staying healthy. There are lots of ways of taking exercise to suit different people. When you exercise, different parts of your body respond in order to keep your muscles working. Think about the changes you feel when you start exercising.

10.1 What is exercise?

Exercise improves your body in a number of different ways. It can make your heart more efficient, it can strengthen your muscles, it can stop you becoming obese and can improve your flexibility and balance. There are many different ways to exercise, but all types of exercise have one thing in common. They involve moving parts of your body and this means making your muscles work. When you move, one of your muscles has to contract and pull on a bone in your skeleton.

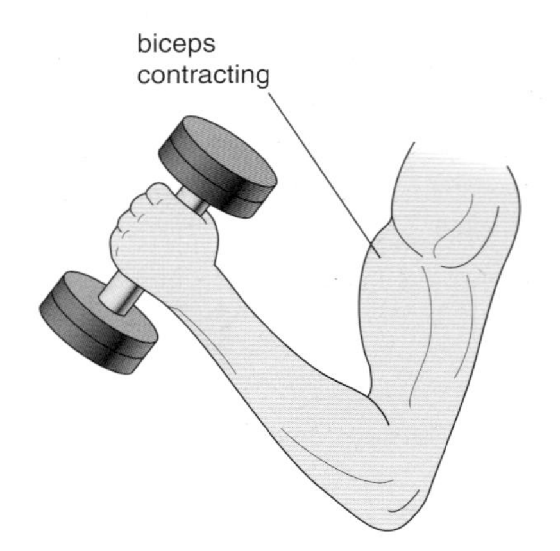

Figure 10.2 The biceps muscle has to contract to pull the arm upwards.

A muscle needs energy whenever it contracts. This energy is released during **respiration** from nutrient molecules derived from the food we eat. Respiration takes place in almost all the cells in your body. It is particularly important in muscle cells.

Respiration is a series of chemical reactions that release energy from glucose and other nutrients. It takes place in the cells of living organisms.

Aerobic respiration requires oxygen to release energy from nutrients.

Anaerobic respiration releases energy from nutrients without oxygen.

In this chapter, you will study two types of respiration: **aerobic** respiration and **anaerobic** respiration. Aerobic respiration (Section 7.3) is the most efficient type of respiration that uses oxygen to release the energy from nutrients. The most important nutrient used in respiration is glucose, a type of sugar. Anaerobic respiration is less efficient, but it does not need oxygen to proceed. Both types of respiration are really a series of chemical reactions, although each one can be summarised by a single equation.

Aerobic respiration

Glucose + oxygen → carbon dioxide + water (plus the release of energy)

Anaerobic respiration

Glucose → lactic acid (plus the release of energy)

The glucose and oxygen used during respiration have to be transported to the muscle cells for respiration to take place. At the same time, carbon dioxide produced during respiration has to be removed from the cells and excreted. Your body has a transport system that it relies on to do these jobs. We will study this in the next section.

1. Choose three pictures from Figure 10.1. For each one, write down the muscles that are contracting during the exercise shown.
2. What substances do the muscles need to obtain energy?
3. How do these substances get into the body?

How are substances transported around your body?

The **circulation system** comprises the heart, blood vessels (arteries, veins, capillaries) and the blood.

In Chapter 9 we learnt how oxygen, carbon dioxide and glucose move into and out of the blood by diffusion. We also studied how active transport can be used to move glucose against a concentration gradient when it is absorbed into the bloodstream from the small intestines. Once substances are absorbed into blood in one part of your body, they are transported to other parts by your **circulation system**.

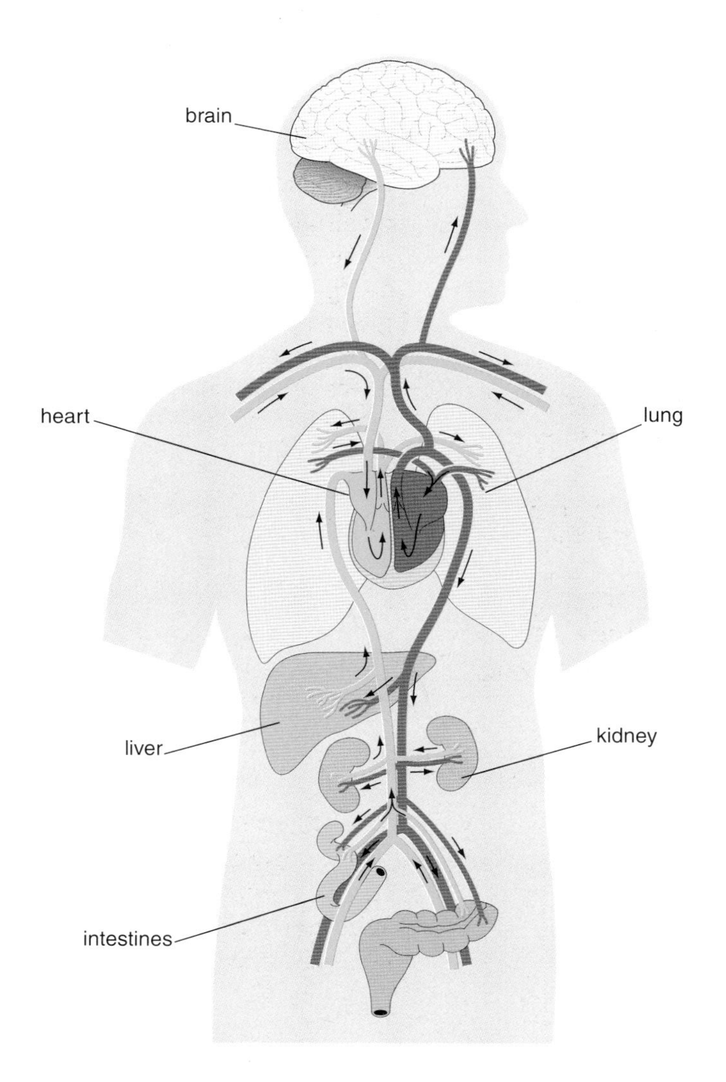

Figure 10.3 The circulation system showing the main organs with which the blood exchanges substances (key: blue = deoxygenated blood; red = oxygenated blood)

The organs in your body would not obtain any of the substances they need without the circulation system, nor would they be able to get rid of any waste products. The circulation system is composed of:

Arteries are blood vessels that carry blood away from the heart to other body organs.

Capillaries are very small blood vessels that carry blood through body organs.

Veins are blood vessels that carry blood back to the heart from other body organs.

Humans have a **double circulation system** with two loops: one from the heart to the lungs and back, and another from the heart to the rest of the body and back.

4 Why is the human circulation system sometimes referred to as two separate systems?

- the heart, which pumps blood around the body;
- **arteries**, which carry blood away from the heart to organs in the body;
- **capillaries**, which carry blood through organs;
- **veins**, which carry blood back to the heart;
- blood, which carries substances around the body.

The heart

The heart is made of special muscle tissue called cardiac muscle. Unlike the muscles that move your bones, cardiac muscle does not require an electrical impulse from a nerve to make it contract. Instead, it contains specialised cells called pacemaker cells that cause each contraction. Every time the heart contracts, it pumps blood around the body. The heart is divided down the middle into two halves. Each half is split into two chambers. The upper chamber in each half is called an atrium and the lower chamber is called a ventricle.

The heart's right atrium fills with blood that has returned from the rest of the body along the vena cava. The blood is pumped into the right ventricle and from the ventricle to the lungs. Here the blood picks up oxygen and gets rid of carbon dioxide. After this, the blood returns to the heart via the left atrium. Then, it flows into the left ventricle before being pumped around the body along the aorta. The system is often referred to as a **double circulation system**, or even as two separate systems – one serving the lungs, the other the rest of the body.

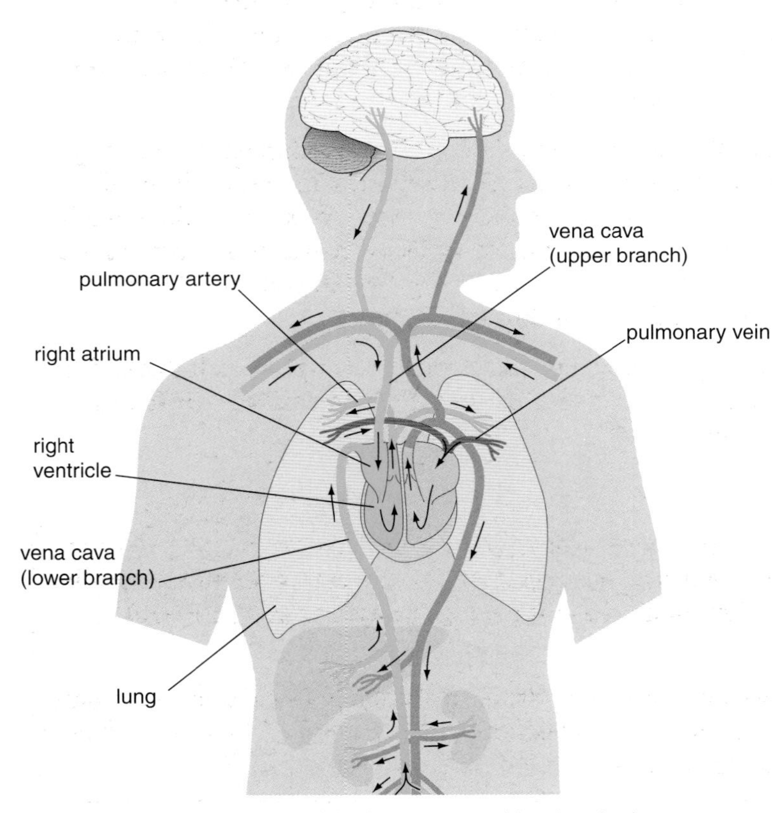

Figure 10.4 The right side of the heart pumps blood to the lungs to absorb oxygen and dispose of carbon dioxide.

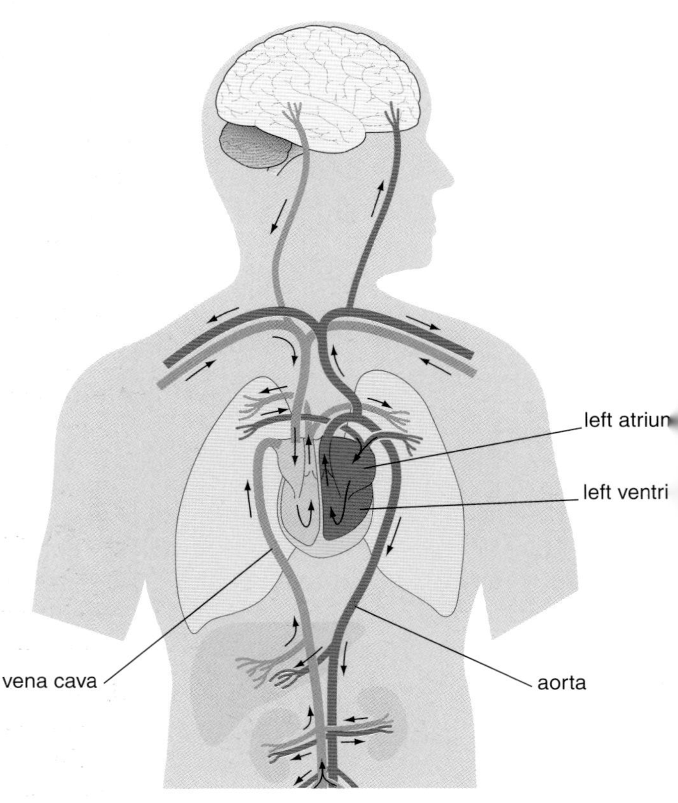

Figure 10.5 The left side of the heart pumps oxygenated blood to rest of the body.

Blood vessels

There are three types of blood vessel: arteries, veins and capillaries. They are the tubes that carry blood around your body as part of your circulation system.

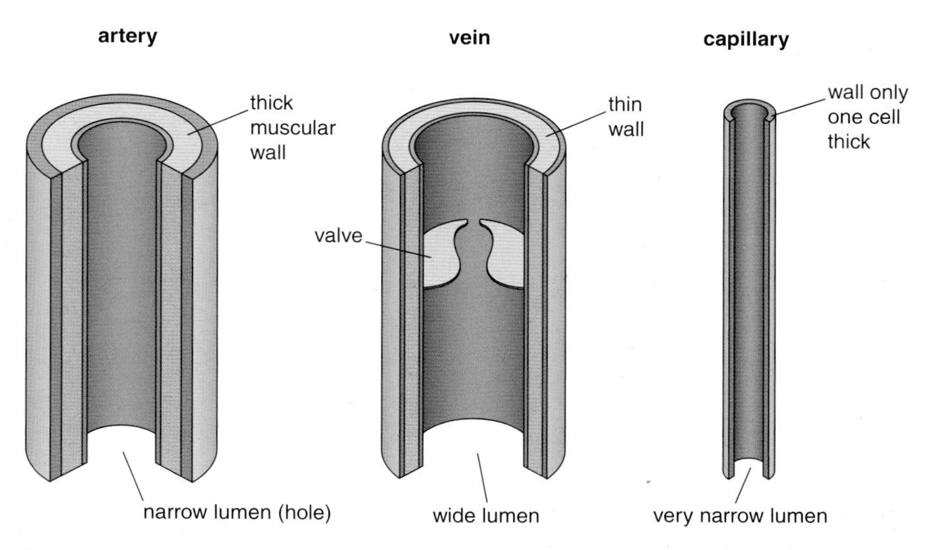

Figure 10.6 Arteries, veins and capillaries have very different structures.

Arteries are blood vessels that carry blood away from the heart. The blood is under pressure, so the arteries need thick walls. The thick muscle in the artery wall contracts to pump the blood. This keeps the blood moving to the organs after it has left the heart.

Veins are wider than arteries, but they have much thinner walls. They collect the blood as it drains from the capillaries and return it to the heart. The blood is not under pressure so veins do not need thick walls like arteries. However, most veins carry blood upwards against gravity in the body. The valves you can see in Figure10.6 stop the blood flowing in the wrong direction through a vein.

Capillaries are very narrow and their walls are only one cell thick. This allows substances to pass in and out of the blood easily by diffusing through the capillary wall.

5. Describe the pathway of blood as it flows around the body. Try to include the names of the important blood vessels through which it flows and the main organs it passes through. Start at the right atrium.
6. What substances pass into or out of the blood in each of the organs you have mentioned in question 5? You may need to look back at Chapter 9 for help.

7 Copy and complete Table 10.1 to summarise how the structures of the different blood vessels are linked to their function. The first one has been done for you.

Blood vessel	Features	How is the feature linked to the vessel's function?
Artery	Thick wall	Carries blood under high pressure
Capillary		
Vein		

Table 10.1 A summary of the structure and function of different blood vessels

10.3 How does blood carry different substances?

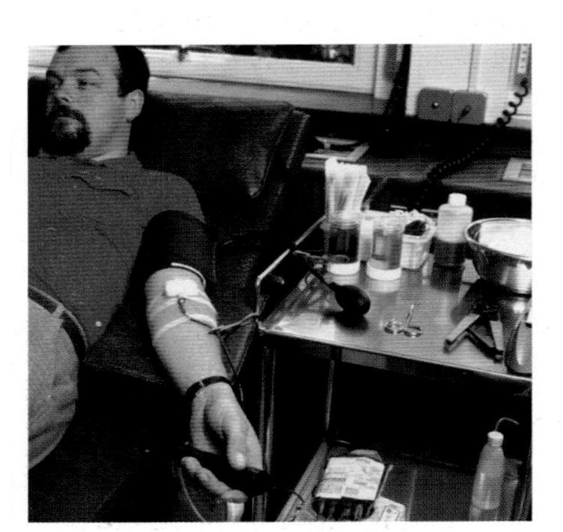

Figure 10.7 This person is donating blood that will be used in a blood transfusion. Do you know what blood is made up from?

Although blood just looks like a red liquid it is, in fact, a mixture made up of four parts. These are:

- blood **plasma**;
- **red blood cells**;
- white blood cells;
- **platelets**.

Plasma and red blood cells carry different substances around the body. White blood cells destroy microorganisms when you get an infection. Platelets help your blood to clot when you cut yourself.

Blood plasma

Blood plasma is mostly water. Soluble substances are dissolved in this water and transported around the body. When the substances reach parts of the body where they are used, they pass out of the plasma and through the capillary walls.

Carbon dioxide is carried in the blood plasma. It diffuses out of cells, where it is produced during respiration, and passes through the capillary wall into the plasma. The carbon dioxide dissolves in the plasma and is eventually transported to the lungs (Figure 10.9). Here, it diffuses out of the plasma and into the alveoli (Section 9.2). The carbon dioxide is then breathed out of the lungs and into the air.

When our food is digested, proteins, fats and carbohydrates are broken down to form much smaller nutrient molecules such as glucose and amino acids. These small, soluble molecules are absorbed from the small intestine into the blood, where they dissolve in plasma. As you know, glucose is needed in your cells for respiration. Blood plasma transports the glucose and other nutrient molecules from the small intestine to your cells so that respiration can take place.

Plasma is the clear watery liquid in blood. It transports soluble substances such as carbon dioxide and nutrients.

Red blood cells are specialised cells that carry oxygen.

Platelets are cells suspended in the blood plasma.

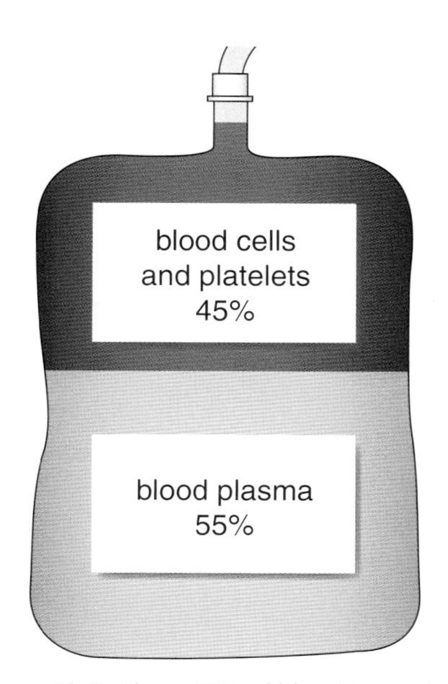

Figure 10.8 About 55% of blood is a pale yellow liquid called blood plasma. The other 45% is made up of red and white blood cells and platelets.

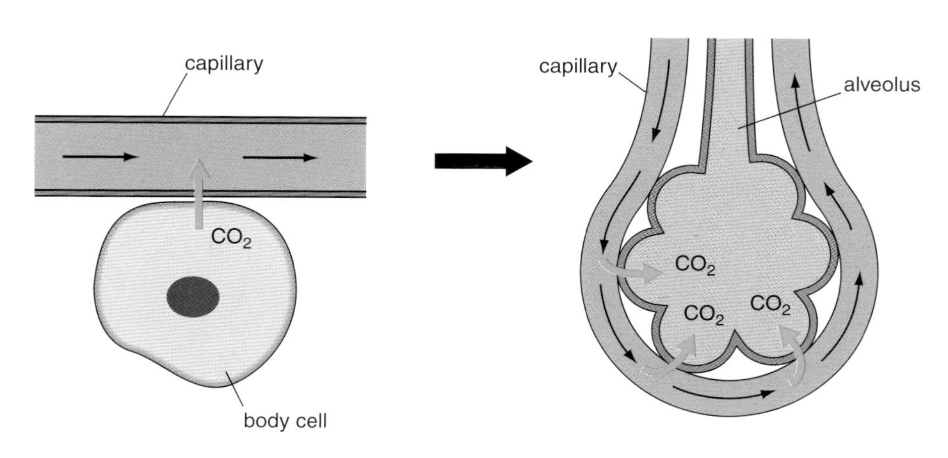

Figure 10.9 Blood plasma transports carbon dioxide from cells to the lungs.

Amino acids that are not required by the body can become harmful. These amino acids are transported to the liver where they are broken down into a chemical called urea. In Section 9.4 we learnt how urea is removed from the blood in the kidneys. Urea is transported from the liver to the kidneys dissolved in blood plasma.

8. Draw a table to summarise the function of each constituent of blood.
9. Explain why blood plasma is needed for respiration to take place.
10. How is blood plasma involved in maintaining the correct amounts of amino acids in the body?

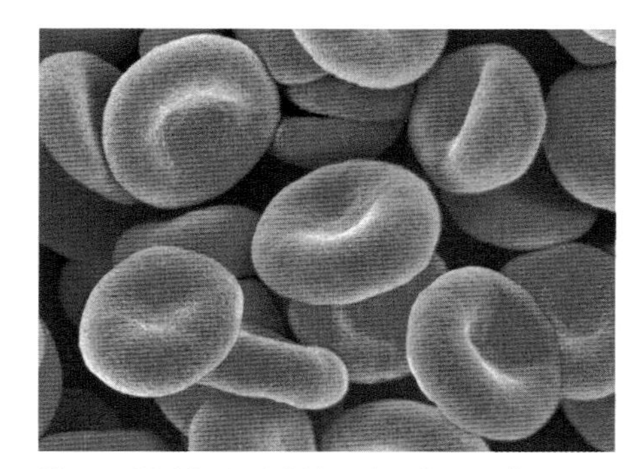

Figure 10.10 A red blood cell is well adapted for transporting oxygen.

Red blood cells

It is crucial that all your cells receive a good supply of oxygen. Oxygen can dissolve in plasma, but this way of transporting oxygen would not supply sufficient oxygen to keep all your cells functioning. However, red blood cells have special features that make them incredibly efficient at transporting oxygen.

- They have a large surface area due to their 'bi-concave' shape.
- They have no nucleus, so that they can carry as much oxygen as possible.
- Most importantly, they are packed with a substance called **haemoglobin**.

Haemoglobin is a substance that can combine with oxygen reversibly. When blood passes through the lungs, oxygen diffuses from the alveoli into the blood (Figure 10.11) where it combines with haemoglobin molecules to form **oxyhaemoglobin**. When the oxyhaemoglobin reaches

Haemoglobin is the chemical in red blood cells that combines with oxygen to form **oxyhaemoglobin**. Oxyhaemoglobin releases the oxygen to cells in the body.

other organs, the oxygen splits from the oxyhaemoglobin and diffuses into the organ's cells, where it is used for respiration. This leaves haemoglobin in the blood that can combine with more oxygen when it next passes through the lungs.

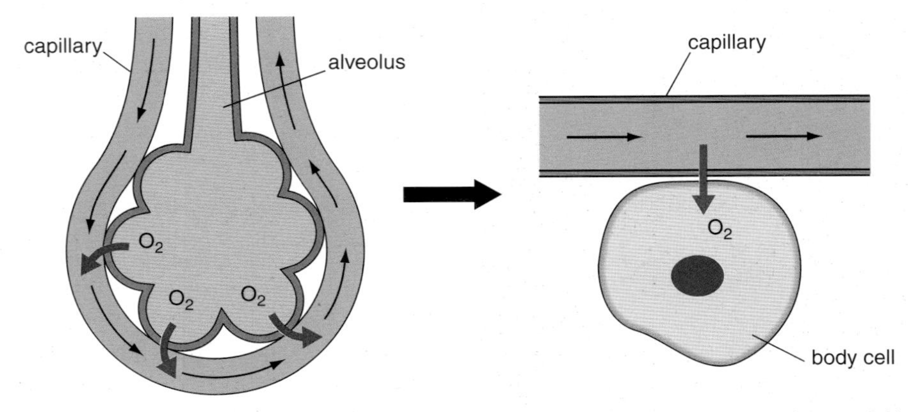

Figure 10.11 Haemoglobin transports oxygen from the lungs to other organs as oxyhaemoglobin in the blood.

Every year, people suffer from carbon monoxide poisoning as a result of faulty gas boilers. Carbon monoxide is a gas that can combine with haemoglobin like oxygen. In fact, carbon monoxide combines about 240 times more strongly with haemoglobin than does oxygen. It forms carboxyhaemoglobin, which is more stable than oxyhaemoglobin and prevents oxygen combining with the same molecules of haemoglobin. This greatly reduces the blood's ability to carry oxygen. As a result, body cells receive insufficient amounts of oxygen, which can lead to unconsciousness and even death. This is why it is so important that gas boilers are serviced regularly.

11. How are red blood cells adapted to absorb and carry oxygen?
12. Our bodies use iron in our food to make haemoglobin. People who have a deficiency of iron in their diet often complain of tiredness. Explain why you think this happens.

How does your body respond when you take exercise?

Figure 10.12 When this runner starts the race, her muscles will need more glucose and oxygen.

So far in this chapter, you have learnt that respiration releases the energy that allows your muscles to contract when you are exercising. You have also learnt how the glucose and oxygen needed for respiration are transported by the circulation system to your body cells. When you exercise, your rate of respiration increases to provide your muscles with enough energy for the extra exertion. This has an immediate effect on the systems involved in providing glucose and oxygen to your muscles.

Look at the runner in Figure 10.12. Her breathing system is providing sufficient oxygen to maintain her body as she waits on the starting line. As soon as the race starts, her muscles will begin contracting more quickly. This needs more energy, so her rate of respiration has to increase. This in turn requires more oxygen and more glucose. Her circulation system will need to work at a rate sufficient to transport this oxygen and the glucose needed to her muscles and other organs.

- When the race starts, the runner's breathing must respond so that more oxygen is absorbed into the blood in her lungs. The number of times she breathes in and out each minute (her breathing rate) will increase. She will also take in more air every time she inhales. On average, the lungs can increase the amount of air inhaled by about eight times when someone starts exercising.
- Her heart rate will also increase when she starts running. This will transport oxygen and glucose to the muscles more quickly.
- To allow more blood to reach the muscles, the arteries supplying the muscles dilate (become wider).
- As the rate of respiration increases, more carbon dioxide is produced. The increased blood flow and increased breathing rate deal with this by removing the extra carbon dioxide as it is produced.

Glycogen – the glucose store

Glycogen is an insoluble polymer formed from glucose and stored in muscles. It can be converted back to glucose and used during vigorous exercise.

Although the blood is being pumped to the muscles more quickly, sufficient glucose may not be absorbed from the small intestine in the first place. When this happens, the muscles use some of their stored **glycogen**. Muscles can store any excess glucose they receive as glycogen. Then, when respiration increases, the muscles convert this glycogen back to glucose so that energy can be released from it. This will help a runner during a race, but it can take two days to replace the glycogen that is used up.

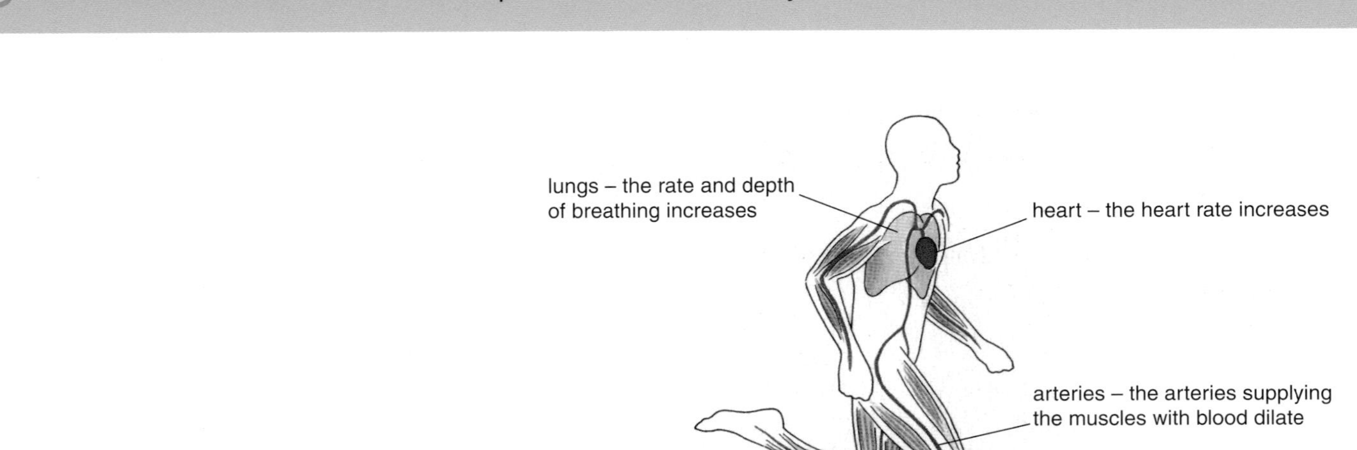

Figure 10.13 When you exercise, different parts of your body respond to increase the rate at which energy is released in your muscles.

13. Imagine you have just set off on a bike ride.
    a) Describe the changes that take place in your body as you start to exercise.
    b) Explain how each of the changes helps your muscles to continue pedalling your bike.
14. How do you think smoking would affect the body's ability to respond to exercise?

Activity – Measuring the effects of exercise on the body

Sara and Kerry decided to investigate the effect of exercise on their heart rate and breathing rate. They used data loggers to measure their heart and breathing rates at rest and during exercise. Their heart rate was measured in beats per minute and their breathing rate in breaths per minute. Their results are shown in Table 10.2.

1. Why did they measure their heart and breathing rates at rest?
2. a) One of Sara's results does not fit the overall pattern. Which result is this?

	Sara		Kerry	
	Heart rate in beats per minute	Breathing rate in breaths per minute	Heart rate in beats per minute	Breathing rate in breaths per minute
Resting	68	12	72	12
Walking	63	12	86	12
Running on the flat	112	16	120	14
Running up hill	126	19	134	17

Table 10.2 Sara and Kerry's results

b) What could have caused this anomalous result?

3. The first two breathing rates are the same for both girls. Why do you think their breathing rates did not increase when they started walking?
4. Describe and explain the general pattern in each student's results using the scientific ideas covered in this chapter.

When Sara and Kerry had stopped running, they measured their heart rate every minute until it returned to the normal heart rate at rest. The time it takes to do this is called your recovery time. Recovery time gives an indication of fitness – shorter recovery times are linked to higher levels of fitness. Their results for this part of the investigation are shown in Table10.3.

5. Plot both students' results as line graphs on one set of axes.
6. a) What was the recovery time for each student?

Time in minutes	Sara	Kerry
0	126	134
1	115	128
2	107	121
3	95	112
4	83	105
5	74	96
6	68	88
7	68	80
8	68	72
9	68	72

Table 10.3 Sara and Kerry's results for recovery time

b) Which student does this suggest is the fitter?

c) What other factors should be considered before drawing a firm conclusion from these data?

10.5 What are the effects of very vigorous exercise?

In Section 10.4, we came to the conclusion that our muscles can keep working provided they have a sufficient supply of oxygen and glucose.

Figure 10.14 Exercising vigorously for long periods of time can put your muscles under great stress.

Even when the supply of glucose is limited, our muscles convert glycogen into glucose in order to maintain respiration. However, during long periods of vigorous exercise this situation can change.

Muscle fatigue occurs during long periods of vigorous exercise and causes the muscles to work less efficiently.

The person in Figure 10.14 is exercising very hard. If the exercise continues for a long time his muscles will start to suffer from a condition called **muscle fatigue**. This occurs when the blood cannot supply the muscles with enough oxygen. When this happens, the muscles start to respire anaerobically.

Lactic acid and oxygen debt

Lactic acid builds up in muscles during anaerobic respiration.

During anaerobic respiration, energy is released from glucose without oxygen. However, anaerobic respiration releases far less energy per gram of glucose than aerobic respiration. This is because anaerobic respiration only partly breaks down glucose, producing a chemical called **lactic acid**. As lactic acid builds up in our muscles, it prevents them contracting properly. This is when muscle fatigue sets in. Just hold your arms straight out in front of you for a long time if you want to experience muscle fatigue!

Figure 10.15 These sprinters have produced a lot of lactic acid in their muscles during the race. They are suffering from muscle fatigue.

During a sprint, athletes expend so much energy so quickly that they must rely on anaerobic respiration. Their muscles have to work so hard that the blood simply cannot supply their muscles with oxygen quickly enough. This is possible for the duration of the race, but at the end of the race their muscles will contain a lot of lactic acid. This must be broken down otherwise it will start to damage their muscles. Lactic acid is oxidised by oxygen to carbon dioxide and water. The carbon dioxide is then excreted as if it had been produced by aerobic respiration. The amount of oxygen needed to oxidise the lactic acid is called an **oxygen debt**. The sprinters in Figure10.15 are paying back their oxygen debt.

An **oxygen debt** is the amount of oxygen needed to oxidise the lactic acid that has built up in muscles during anaerobic respiration.

15 a) 'Anaerobic respiration is far less efficient than aerobic respiration.' Explain what this statement means.
b) How does your body respond when you cycle up a steep hill quickly?
c) Once you finish cycling up the hill, you need a rest. Explain what happens in your muscles while you are resting. Use all the key words in the definition boxes on the left in your answer.

Summary

- ✓ **Aerobic** respiration releases energy from glucose using oxygen. **Anaerobic** respiration releases much less energy than aerobic respiration, but it does not require oxygen.
- ✓ Substances are absorbed into blood in one part of your body and transported to other parts by your **circulation system.**
- ✓ **Arteries** carry blood away from the heart under pressure.
- ✓ **Capillaries** carry blood through organs where substances pass into or out of the blood.
- ✓ **Veins** carry the blood from the body back to the heart.
- ✓ There is a **double circulation** system in humans. One carries blood to the lungs where it absorbs oxygen. The other transports blood to other organs in the body.
- ✓ Blood is made up of **plasma**, **red blood cells**, white blood cells and **platelets**.
- ✓ Blood plasma transports carbon dioxide, soluble nutrients and urea.
- ✓ Red blood cells have specialised features that make them very efficient at transporting oxygen.
- ✓ **Haemoglobin** is a substance in red blood cells that combines with oxygen to form **oxyhaemoglobin**. Oxyhaemoglobin readily releases its oxygen to cells in the body for aerobic respiration.
- ✓ Muscles use the energy released during respiration to contract.
- ✓ During exercise, the heart rate and the rate and depth of breathing increase. In addition, the arteries supplying muscles with blood dilate to increase blood flow.
- ✓ Muscles convert stored **glycogen** into glucose during exercise.
- ✓ Muscles contract less efficiently during long periods of vigorous exercise, because less energy is released from each gram of glucose during anaerobic respiration. A condition called **muscle fatigue** results.
- ✓ During anaerobic respiration, **lactic acid** is produced. This must be oxidised by oxygen after the exercise has finished. The amount of oxygen needed to do this is called an **oxygen debt**.

EXAMQUESTIONS

1

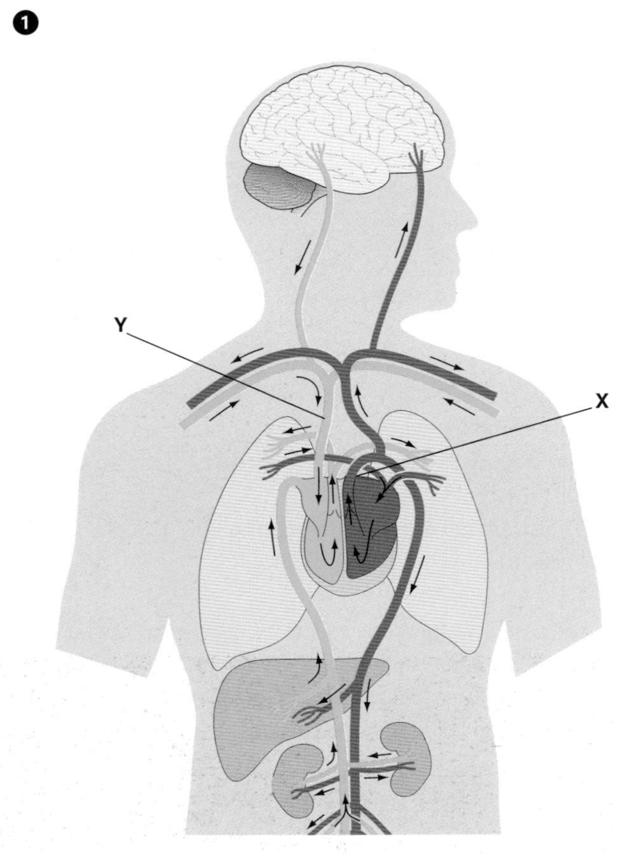

Figure 10.16 The human circulation system

a) What type of blood vessels are X and Y? *(2 marks)*

b) The walls of blood capillaries are only one cell thick. What process occurs more efficiently because of this feature? *(1 mark)*

c) Blood plasma is the watery liquid in blood. Name two substances that are transported in blood plasma. *(2 marks)*

2 Red blood cells are specialised cells that transport oxygen around the body.

a) Describe two special features of a red blood cell. *(2 marks)*

b) Explain how red blood cells transport oxygen. *(2 marks)*

c) Explain how oxygen is released from a red blood cell where it is needed. *(2 marks)*

3 When athletes start exercising, their bodies respond.

a) State two of these responses that increase blood flow to the muscles. *(2 marks)*

b) Explain why blood flow to the muscles must increase when a person starts exercising. *(3 marks)*

c) What is the role of glycogen during exercise? *(2 marks)*

4 During vigorous exercise, aerobic respiration may be replaced by anaerobic respiration.

a) State two differences between aerobic and anaerobic respiration. *(2 marks)*

b) What happens to muscles during long periods of vigorous exercise? *(1 mark)*

c) During a period of anaerobic respiration, a runner builds up an oxygen debt. Explain what is meant by the term ‘oxygen debt’. *(3 marks)*

Chapter 11
How are microorganisms used to make food and drink?

At the end of this chapter you should:

- ✓ be able to describe the theory of spontaneous generation and the discoveries that led to its being disproved;
- ✓ be able to describe the theory of biogenesis;
- ✓ know that high temperatures destroy microorganisms and understand pasteurisation;
- ✓ understand the role of yeast in making bread and alcoholic drinks;
- ✓ be able to describe the process of making beer;
- ✓ know the carbohydrate sources for both beer and wine production;
- ✓ be able to describe the process of yoghurt and cheese manufacture;
- ✓ understand the need for hygiene and pure cultures of microorganisms when making food products;
- ✓ be aware of how to work safely when using microorganisms.

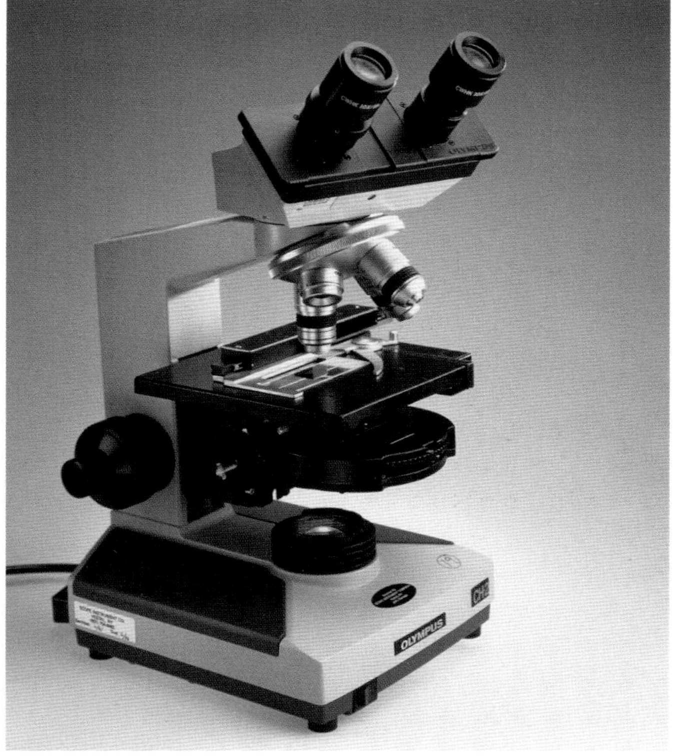

Figure 11.1 Robert Hooke's microscope (left), which he used in the 17th century, and (right) the microscopes you use today in the school laboratory. Notice the differences between the two instruments.

11.1 The theories of spontaneous generation and biogenesis

The **theory of biogenesis** states that all living organisms arise from pre-existing living organisms.

From what you know about cats and kittens, dogs and puppies, the **theory of biogenesis** probably seems obvious to you. The theory states that all living things arise from pre-existing living things. However, this has not always seemed so obvious. In the 4th century BC, Aristotle and other philosophers believed that simple living organisms came into existence from non-living materials by the force of nature. This was called 'spontaneous generation'.

Figure 11.2 A popular view of spontaneous generation in the Middle Ages

It wasn't until the 17th century that people began to question this idea. William Harvey suggested that maggots in rotting meat came from eggs that flies had laid and Francesco Redi developed this idea in 1668. He observed flies settling on food and laying eggs, which then developed into maggots. His ideas provided an alternative explanation that opposed the idea of spontaneous generation. However, many people in the late Middle Ages still believed that insects, and even frogs and mice, could be created from, say, dirty clothes (Figure 11.2).

In the 17th and 18th centuries, the development of microscopes opened up a new world of organisms and structures that had never been seen before (see Chapter 5). With lenses that gave a remarkable magnification of about 300 times, Anton van Leeuwenhoek was able to see 'animalcules' (little animals) when he looked at scrapings from his teeth and samples of pond water. Today we would call these microorganisms, or bacteria.

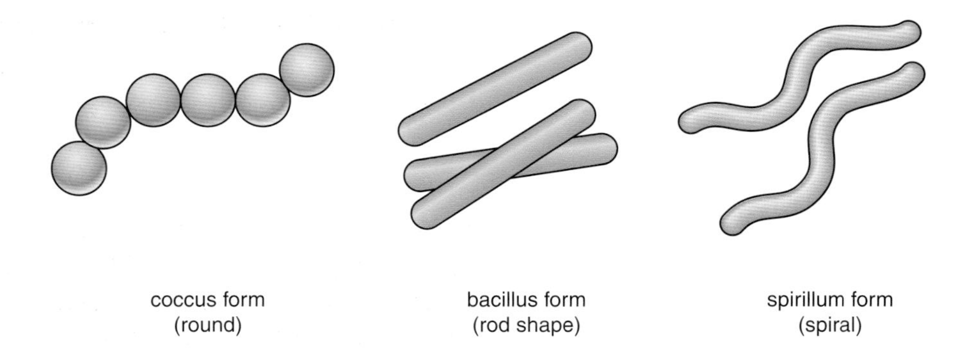

Figure 11.3 The different shapes of bacteria – spherical, rod-shaped and spiral

Spallanzani, Schwann and Pasteur: evidence against spontaneous generation

In the 1760s Lazzaro Spallanzani was one of the first people to investigate the concept of biogenesis scientifically. Spallanzani said that when using a microscope he had seen microbes move. Therefore these microbes were living organisms, like flies or frogs. Could microbes be produced spontaneously? Spallanzani knew about experiments with flasks of meat broth that went 'off' and showed microorganism growth, even if the flask was sealed.

Spallanzani said that these microbes were already present in the broth. He carried out experiments to show that no microorganisms would grow if the flask were first boiled for a long time (Figure 11.4). Theodor Schwann also did experiments to show that yeast cells were capable of increasing in number and caused fermentation only when they were alive. If they were killed by boiling, no alcohol was formed.

However, the arguments about spontaneous generation continued well into the 19th century. Some critics argued that Spallanzani's experiments did not disprove spontaneous generation because he sealed his flasks. These critics thought air contained the 'active principle' for spontaneous generation and so had to enter the flasks for life to be created from non-living materials. So, in 1860, the Paris Academy of Sciences offered a prize for anyone who could put an end to the long debate on spontaneous generation. The prize was claimed by Louis Pasteur, who showed that normal air by itself did not allow microorganisms to grow. His clever apparatus design allowed air but not bacteria to enter the flask (Figure 11.5).

1. What is the main difference between the theory of spontaneous generation and the theory of biogenesis?
2. Leeuwenhoek noticed that there were fewer 'animalcules' present in his tooth-scrapings after a hot drink or rinsing his mouth with vinegar. Suggest an explanation for his observations.

Cell theory: a mechanism for biogenesis

In 1839 Schwann introduced our modern theory of cells by stating that cells are separate units with their own cellular chemistry (Chapter 5). Together with Matthias Schleiden, he identified and produced diagrams showing membranes and nuclei, and recorded features that were common to plant and animal cells. Today, we know that cells are the basic units of living organisms. We also know that all cells are formed from other cells by cell division (mitosis or meiosis). All living organisms come from 'parent' organisms by either asexual or sexual reproduction. This is the explanation for the theory of biogenesis.

Although we might find the idea of spontaneous generation ridiculous today, we should not despise earlier scientists. The idea of spontaneous generation was a reasonable attempt to explain what people saw. The theory of biogenesis, which replaced spontaneous generation, required the development of a scientific method of investigation.

Activity – Following the experiments of Spallanzani and Pasteur

Examine carefully the techniques that Spallanzani and Pasteur used in the following experiments. Both men followed scientific methods by:

i) identifying a problem;
ii) drawing up a hypothesis to test;
iii) performing an experiment;
iv) using the evidence of results to draw a conclusion.

Keep in mind your knowledge of fair tests and control variables.

Spallanzani used the apparatus shown in Figure 11.4.

1. Suggest a possible hypothesis for Spallanzani to test.
2. The key point about Spallanzani's experiment was that he boiled the broth in the flask (for up to 45 minutes). Previous scientists had

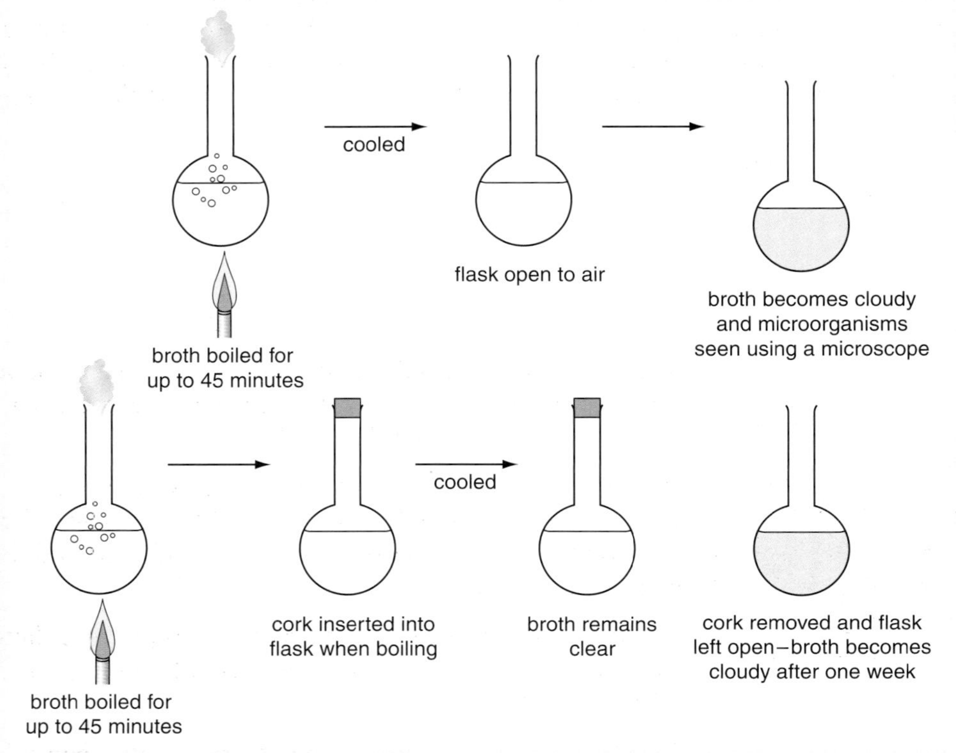

Figure 11.4 A flow diagram showing Spallanzani's experiment

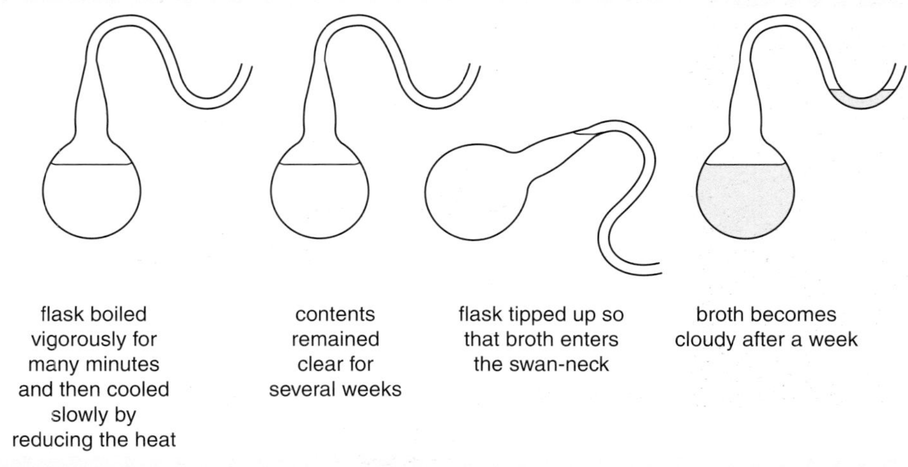

Figure 11.5 Pasteur's experiment using a swan-necked flask

used 'washed-clean' flasks into which they poured the broth after heating for a short while. What difference do you think this made to Spallanzani's results?

3. What factors (possible variables) were kept constant during the experiment?
4. What could Spallanzani conclude about spontaneous generation from the closed flask?
5. What do you think the open flask showed?

Louis Pasteur had already shown that problems experienced by the brewing industry were caused by living microorganisms. Pasteur knew about the criticisms of Spallanzani's experiments (see page 179). Pasteur devised an ingenious 'swan-necked' flask. He heated the broth in his swan-necked flask and then cooled it slowly so that air could enter the flask but any bacteria entering on dust particles would settle out in the bottom of the neck without rising up and entering the broth (Figure 11.5).

6. How did Pasteur's flask overcome the criticisms of Spallanzani's experiment?
7. Was there any evidence of cloudiness in the broth in the swan-necked flasks after standing for several days?
8. What happened a few days after the broth was tipped into the neck of the flask and then back into the body of the flask? Suggest a reason for the change.

11.2 Yeast – a very useful microorganism

Yeast is a single-celled microorganism that occurs naturally on fruits, grains and even cabbage leaves. It appears as a greyish bloom on the surface of the plant (see Figure 11.14).

People have used yeast for thousands of years; ancient civilisations used it to produce bread and alcoholic drinks. Water was frequently contaminated, so alcoholic drinks were safer to drink because bacteria could not survive in them.

Yeast reproduces asexually by budding to produce daughter cells. The daughter cells grow, then separate off and repeat the process (Figure 11.6).

Yeast cells respire, using oxygen from the air when it is available. This is aerobic respiration (Chapter 10). The products of aerobic respiration are carbon dioxide and water.

3. What does a yeast cell have in common with a plant cell?
4. Explain why no alcohol is produced after boiling a solution containing yeast.

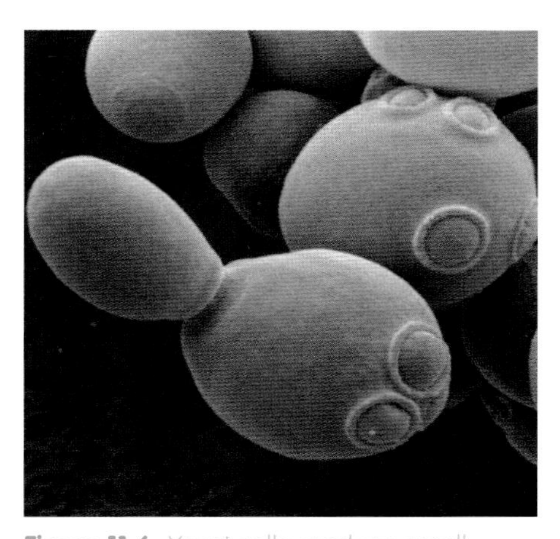

Figure 11.6 Yeast cells produce small daughter cells growing on the side. This is called 'budding'.

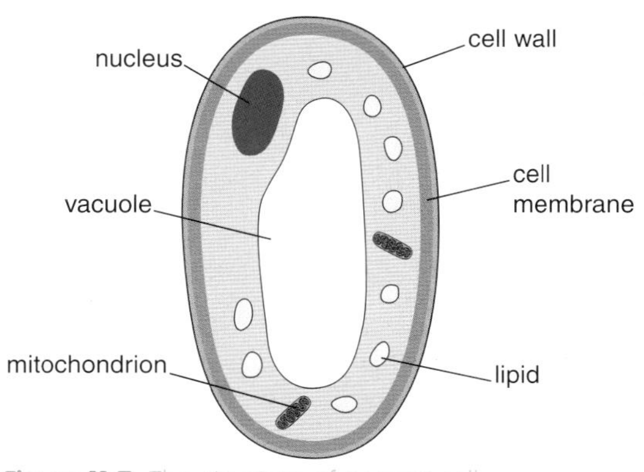

Figure 11.7 The structure of a yeast cell

Fermentation is a type of anaerobic respiration by microorganisms, using enzymes to break down glucose to produce carbon dioxide and ethanol.

sugar (carbohydrate)	+	oxygen	→	carbon dioxide	+	water
$C_{12}H_{22}O_{11}$	+	$12O_2$	→	$12CO_2$	+	$11H_2O$

Yeast can also respire anaerobically – that is, without oxygen. This process is called **fermentation**. The products are carbon dioxide and alcohol (ethanol):

sugar	+	water	→	ethanol	+	carbon dioxide
$C_{12}H_{22}O_{11}$	+	H_2O	→	$4C_2H_5OH$	+	$4CO_2$

Anaerobic respiration produces less cellular energy than aerobic respiration. So yeast cells need to respire aerobically to release the most energy to grow and reproduce.

5. Fermented foods and drinks have been made by many societies. What was the advantage of using fermentation processes to people long ago?

Activity – Investigating natural yeasts

SAFETY: Although you are conducting this investigation with everyday foods, you must not eat or drink any items that you use or produce in the laboratory. These products will not be pure and could be contaminated with bacteria or traces of chemicals that are toxic.

In this investigation, the yeast respires anaerobically.

- First, collect some baker's yeast and three or four of the following: ripe plums, ripe grapes, outer cabbage leaves, grains – or anything else you see with a yeast bloom.
- Set up pairs of tubes as shown in Figure 11.8 for each of the three or four samples chosen.
- Make up a 2% sugar solution (2 g of sugar in 100 cm^3 water)
- Crush the fruits to release the juice, keeping the skins in the pulp while crushing.
- Pour the juice or the baker's yeast into the large tube and then fill it with the sugar solution. Insert the delivery tube so that it is just dipping into the lime water in the small tube.
- Leave all the tubes in a warm environment for a number of days, checking them after one day.

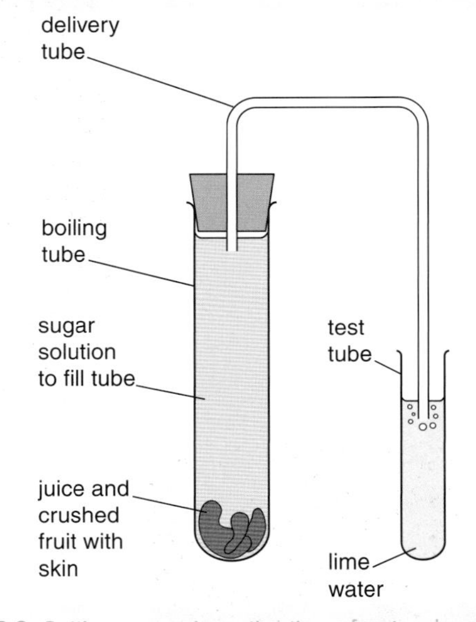

Figure 11.8 Setting up an investigation of natural yeasts

1. What do you notice after about one day?
2. Is there any change in the lime water after a few days? What can you conclude from this?
3. Smell the contents of tube 1. What do you detect?
4. What conclusions can you draw about natural wild yeasts?

Yeast in beer brewing and wine making

Beer and other alcoholic drinks produced by small batch processes date back thousands of years. The production of both beer and wine relies on the anaerobic reaction of yeast with sugar to produce alcohol (ethanol). The large volumes of beer consumed today require bulk industrial production. In industrial brewing, the sequence of processes is automatically controlled. But in smaller, local breweries which make 'real ale', traditional methods are still used. The stages in brewing are as follows.

Malting is the process in which the starch in germinating barley grains is broken down into sugar by enzymes.

1 **Malting:** Barley grains are soaked in water, drained and then incubated at 40 °C to allow them to germinate (Figure 11.9). This activates amylase enzymes in the barley grains, which break down the starch stored in the grains to sugar (Figure 11.10).

Figure 11.9 The process of malting: tossing germinating barley grains in the air

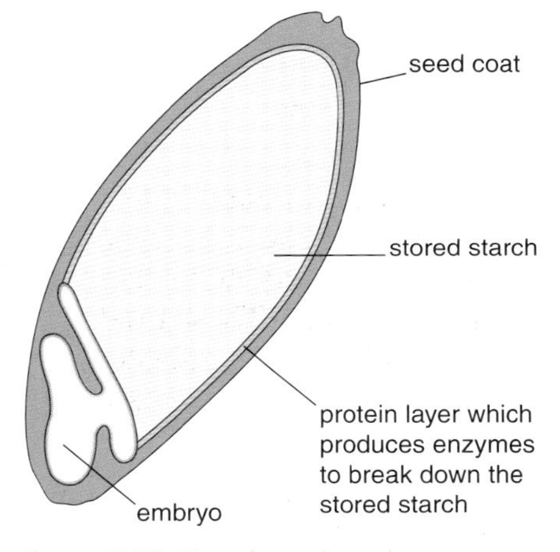

Figure 11.10 Stored starch and enzyme location in the barley grains

2 **Extracting the sugar**: Enzyme action is stopped by heating to 80 °C. The grains are then crushed and the sugar is extracted with hot water.
3 **Flavouring:** The characteristic bitter taste of beer is produced by adding hop flowers (Figure 11.11). In bulk production, resin extracted from the flowers is used. Hops also have a preservative quality.
4 **Fermentation:** This takes place as a batch process (see Section 12.3). Yeast is added to the cooled liquid from the flavouring stage in a fermenter (Figure 11.12). Yeast enzymes convert the sugar to ethanol and carbon dioxide.
5 **Conditioning:** Real ale is conditioned by leaving it to stand in barrels where the remaining yeast settles to the bottom. It ferments very slowly until the alcohol reaches 4–8% and the carbon dioxide dissolves. Fermentation stops when the sugar is used up. After filtering to remove yeast and other materials that cause cloudiness, the beer is bottled or put into kegs.

Figure 11.11 At the base of each petal of an unpollinated hop flower, there is a yellow gland. This produces the resins and oils which give beer its bitter taste.

Figure 11.12 A technician adjusting the fermentation vats in a modern microbrewery

In mass-production breweries, quality control is used to produce the same colour and flavour in each batch. Computers replace the 'master brewer'. They control temperatures, filtering, pasteurising and gas level at various stages of the process. The beer is then packaged into kegs, bottles or cans for sale.

The remains of the grain from the malting stage are not wasted but used as cattle or pig food. In large-scale processes the carbon dioxide produced is converted into solid carbon dioxide.

A large number of scientists from different disciplines are employed in the modern mass production of beer. These include plant geneticists who breed barley with the right amylase enzymes and hops with the desired flavour. Microbiologists produce the ideal yeast cultures (such as *Saccharomyces carlsbergiensis*). Engineers design the brewing equipment. Computer experts programme the brewing processes so that each particular beer always appears and tastes exactly the same.

6. How does the amount of energy differ in the aerobic and anaerobic respiration with yeast?
7. What is the carbohydrate source for making beer?
8. After malting, sugar is extracted from the barley grains at high temperature, producing a sterile solution. What is the significance of this to the final product?

Making wine

Wine making involves the same fermentation reaction as that used to produce beer. However, in this case, wild yeasts are present on the grape skins (Figure 11.14) and the sugars needed for fermentation are inside the grapes.

There are five main stages in wine making.

1 **Harvesting** The grapes are harvested in bunches from the vines either by hand or mechanically.

Figure 11.13 Grapes ready for harvest

Figure 11.14 Notice the natural yeast on the surface of these grapes.

2 **Crushing:** Many years ago, grapes were trampled by foot. However, nowadays mechanical crushers release the sugary juice from the grapes. This reduces contamination by other microorganisms and increases the keeping quality of the wine. Enzymes can be used in wine production to increase the volume and colour of the juice extracted from the grapes, giving a wine with a good fruity taste in less time.

3 **Pressing:** For white wine, the crushed grapes are quickly pressed and the liquid is then removed to a separate vessel. For red wine, the liquid remains in contact with the skins and stem, which give it both colour and flavour from tannins.

4 **Fermentation:** Wild yeasts would naturally begin fermentation at this stage in less than a day. In modern large-scale production, the juice is heated and treated with sulfur dioxide to kill wild yeasts and bacteria before a selected yeast culture is added. Adding a selected yeast culture ensures a predictable product. During the fermentation process, sugar in the grape juice is converted to alcohol. The rise in alcohol concentration inhibits the enzymes and fermentation stops at about 12% alcohol.

5 **Ageing:** The wine is stored in cool conditions either in barrels before bottling or in the bottles. During this stage, different flavours develop in the wine.

To produce a wine that was the same every year, the following factors would have to remain constant: the type of grape, the soil and the amount of rainfall, hours of sunshine, day and night air temperature and the yeast type. But of course these factors vary. Differences in climate and sunlight account for the differences in the sugar content of the grapes. These factors produce variation in the flavour of the wine, and 'good' and 'bad' years for wine production.

9. What is the carbohydrate source for the production of alcohol in wine?
10. Why is modern mass-produced wine heat-treated before adding a selected yeast culture?
11. What environmental factors can increase the sugar content of grapes?

11.4 Yeast in bread making

Pictures on the walls of Egyptian tombs show that bread was a staple food in Ancient Egypt. Bread making probably originated from leftover gruel (wheat grains soaked in water), which had been fermented by wild yeasts on the grain. In different societies, bread has been made from different grains and sometimes from root carbohydrate sources. Selective crop breeding has now produced much larger wheat and hybrid grains, which are ground into flour (Figures 11.15 and 11.16).

Figure 11.15 Emmer wheat used to make bread in ancient times

Figure 11.16 Grains of wheat which are made into flour for use in modern bread making

For bread making, flour with a high gluten content is chosen and this is called 'strong flour'. More gluten protein in the flour makes a more stable loaf with air pockets evenly distributed throughout the loaf.

Both bread making by hand and using a breadmaker follow similar steps.

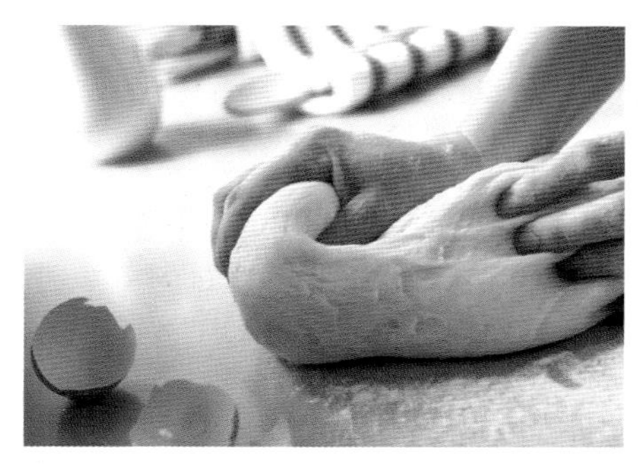

Figure 11.18 Kneading uncooked dough

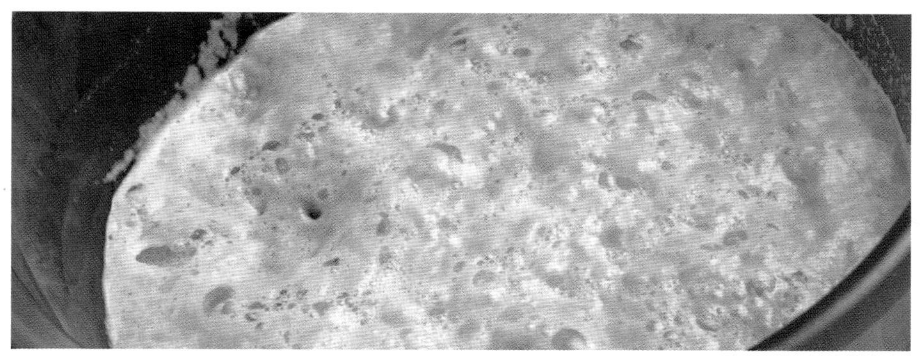

Figure 11.17 Yeast respiring aerobically and producing carbon dioxide

1 **Activation**: The yeast is activated by adding it to warm water with a little sugar. Soon the yeast begins to form bubbles. The yeast is respiring aerobically and the bubbles are carbon dioxide (Figure 11.17).
2 **Kneading:** The activated yeast from stage 1 is added to 'strong' flour to make dough. Then the dough is kneaded until it feels elastic (Figure 11.18). This distributes the yeast uniformly throughout the dough and removes most of the carbon dioxide.

3 **Proving:** The dough is put to one side in a warm place to be 'proved' until it has doubled in size. At this stage, the yeast respires anaerobically and different enzymes in the yeast speed up the breakdown of sugars in the dough to form alcohol and carbon dioxide. The gas produced causes the volume to increase.

The reactions can be summarised in the following word equations:

$$\text{sugar} + \text{water} \xrightarrow{\text{yeast enzymes}} \text{glucose} + \text{fructose}$$

$$\text{glucose} + \text{fructose} \xrightarrow{\text{yeast enzymes}} \text{alcohol} + \text{carbon dioxide}$$

Figure 11.19 Look how much the volume of the dough has increased while proving and leavening.

4 **Leavening:** The dough is kneaded a second time, breaking up the bubbles, and then placed in a warm place. Carbon dioxide produced is trapped in the dough in small bubbles. This is helped by enzymes from the yeast acting on proteins in the dough to make it more elastic. The bubbles expand and increase the volume of the dough (Figure 11.19). This gives the bread its light texture and makes it more easily digested.
5 **Baking:** When the dough has doubled in volume again, it is baked in a very hot oven. The high temperature denatures the yeast enzymes and stops any further reactions. It also evaporates any alcohol and expands the carbon dioxide bubbles. When the loaf is baked, it is removed from the oven and cooled (Figure 11.20).

Figure 11.20 The baked loaf of bread

Today, science and technology are applied to industrial bread making. Wheat has been selectively bred to change the gluten content, and yeasts are genetically engineered to change their enzymes. In the manufacturing process, scientists study how to increase water retention and keep the loaf fresher for longer.

⑫ What is the source of the enzymes for bread making?

⑬ If a thermostat failure caused the proving temperature to rise to 65 °C in a bakery, would you expect to get 'super-sized' loaves? Explain your answer.

⑭ What is the role of gluten protein in the bread-making process?

11.5 Bacteria in yoghurt manufacture

Yoghurt has been made in Turkey for thousands of years from goat's milk stored in goat skins. Yoghurt is a thickened milk product with a sharp taste. The sour taste is produced by bacteria introduced into the milk which break down the milk sugar (lactose) to lactic acid.

The low pH of the lactic acid prevents other microorganisms growing in the milk, and increases its keeping qualities. This was very important for early civilisations without the means to keep milk cool.

Yoghurt production is a batch process. There are five stages in its manufacture.

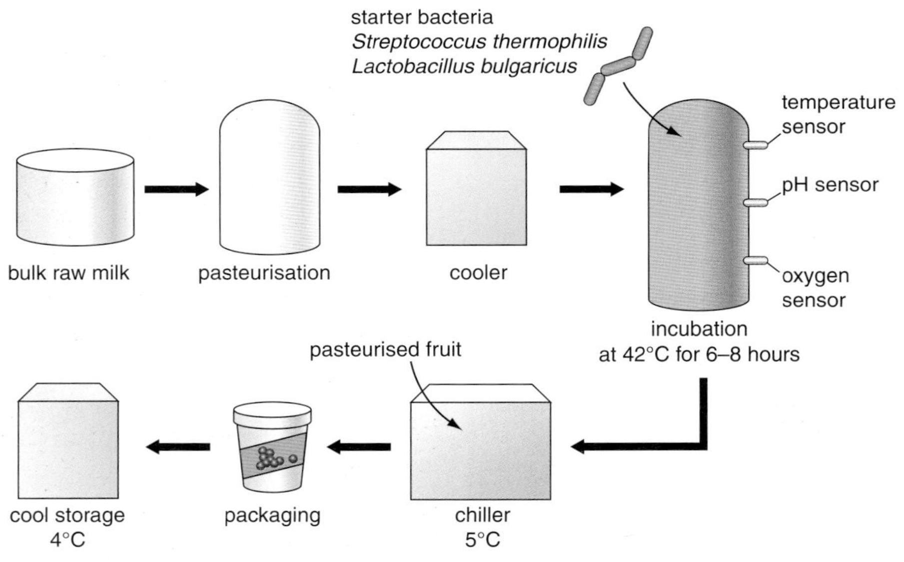

Figure 11.21 A flow diagram for the manufacture of yoghurt

> **Pasteurisation** is the heat treatment process, named after Louis Pasteur, used to kill harmful bacteria.

1 **Pasteurisation**: Yoghurt manufacture starts with raw milk. This is pasteurised by heating the milk to 93 °C. Pasteurisation kills most of the unwanted bacteria.
2 **Cooling:** The pasteurised milk is cooled quickly.
3 **Incubation:** A bacterial culture is added to the cooled milk in an incubation tank. The culture contains *Lactobacillus bulgaricus* and *Streptococcus thermophilus* bacteria (Figure 11.22). These convert

the milk sugar (lactose) to lactic acid. The optimum incubation temperature is 42 °C, which produces the maximum volume of yoghurt in the minimum time (Figure 11.23). As the bacteria respire aerobically, the oxygen level in the mixture is monitored and oxygen is bubbled through it as required. The lactic acid produced lowers the pH to 4.6. This coagulates (thickens) the milk protein (casein), forming a semi-solid. The pH is carefully monitored because if it became too low it would denature the bacterial enzymes. The time taken to make yoghurt in an incubation tank is about 4 to 6 hours.

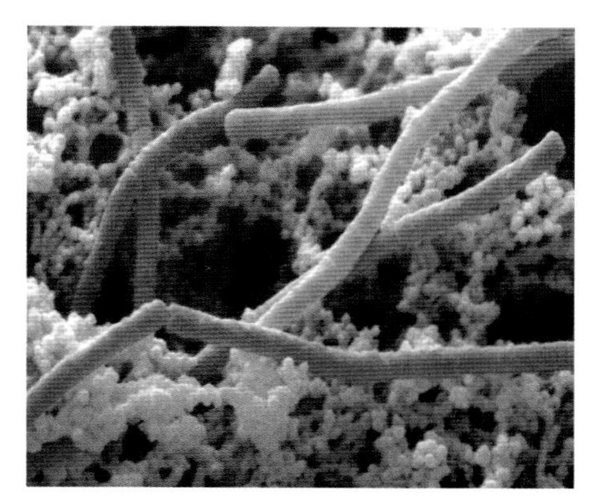

Figure 11.22 *Lactobacillus bulgaricus*, used in cultures for yoghurt manufacture, is a typical rod-shaped bacterium.

Figure 11.23 Incubation tanks used in the manufacture of yoghurt

4 **Chilling:** When the yoghurt has been made, the chiller reduces the temperature to about 5 °C. At this temperature, the bacterial enzymes work very slowly.

5 **Packaging:** Throughout the packaging process, yoghurt pots must be kept in sterile conditions to prevent entry of airborne bacteria or fungal spores. The containers are sealed and stored below 4 °C so that bacterial activity is minimised.

15 At what temperature is the raw milk pasteurised for yoghurt manufacture?

16 If pasteurisation were not carried out, what could happen to the yoghurt?

17 Write a word equation for the chemical reaction that causes the pH of yoghurt to drop.

18 Suppose you have noticed that the foil lid on a pot of yoghurt is damaged. When you open the yoghurt there is a strange smell. Explain its possible cause.

19 a) How did Pasteur's careful investigations, described in Section 11.1, help both the beer and wine industries?
b) Why should you remember Pasteur when you drink milk or eat yoghurt?

11.6 Bacteria in cheese making

Cheese making uses bacteria that metabolise at a lower temperature than those used for making yoghurt. In the past, milk kept in a cool dairy was sometimes contaminated with bacterial spores and the result was cheese. When cheese is made today, rennet is used to coagulate the milk and enzymes speed up the ripening process.

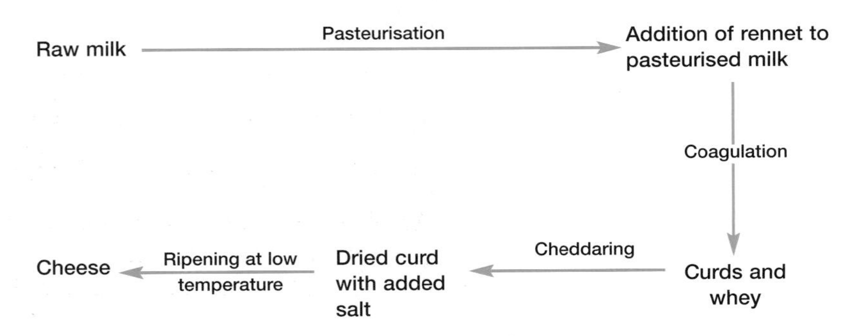

Figure 11.24 A flow diagram to illustrate how cheese is made

There are four main stages in cheese making.

1. **Pasteurisation**: The cheese-making process begins in the same way as yoghurt manufacture with pasteurisation of the milk.
2. **Coagulation:** A culture of lactic acid-producing bacteria is added to convert the lactose to lactic acid. Rennet, which contains an enzyme to coagulate the milk protein, is then added (Figure 11.25). This results in the production of solid curds and liquid whey.
3. **Cheddaring:** To make a hard cheese such as cheddar, the curd is chopped, drained and salted (Figure 11.26). At the end of this stage, the pH has dropped to about 5. This low pH together with the added salt act as preservatives. The cheese is then cut into blocks and coated with wax to keep out oxygen and unwanted microorganisms.
4. **Ripening:** The blocks of cheese are ripened at a low temperature so that the enzymes can break down fats and proteins, giving the cheese its familiar taste when ripe (Figure 11.27).

Figure 11.25 Adding rennet to coagulate the milk protein (casein)

Figure 11.26 Cheddaring

Figure 11.27 Cheeses ripening in a cool room

Cheese keeps well, but not for very long periods because some moulds can grow at the low pH in cheese even at refrigerator temperatures.

Blue cheeses are made by adding mould spores after the coagulation stage. As the fungal hyphae respire aerobically, they do not develop until holes are made by inserting wire in the cheese at the ripening stage. Air can now enter where the wires have been and the spores germinate. As the spores form hyphae, blue veins develop around the air holes.

20 What factors or materials act as a preservative in cheese?

21 Why does cheese take longer to make than yoghurt?

22 Why do the veins in 'blue cheese' not develop until holes have been made in the cheese?

How can we be sure to use microorganisms safely?

As we have seen, the production of beer, wine, yoghurt and cheese using microorganisms gave people safer, longer-lasting food sources before the age of preservatives or refrigeration. But they still had problems to face. Among the many useful species of microorganisms there are, in addition, those that cause putrefaction (food spoiling by microorganisms). Wild yeasts are present on the skins of fruits and the yeast cells can blow around in the air. Bacteria are also carried in the air. Contamination by these microorganisms can result in the production of toxic or foul-tasting substances that can cause illness or result in food that cannot be eaten.

How is the problem of 'wild' microorganisms overcome?

In the 19th century, Pasteur found that, to prevent wine and beer souring, the drinks had to be heated gently to between 50 and 55 °C for several minutes. This process is called pasteurisation (Section 11.5). Most dairy products are still treated in this way. The times and temperatures used for pasteurisation are different for each process, but must be high enough to kill spoiling or pathogenic bacteria without changing the taste of the product.

In the modern food and drink industry, several other steps are taken to avoid contamination. These precautions mean that you can safely eat and drink these products provided the packaging is not damaged.

- When bacterial cultures are needed, pure strains of bacteria and yeasts are used. These strains are carefully selected and may be genetically engineered to produce a final product quickly and with the desired qualities. If the strain is pure, it is uncontaminated by other microorganisms.

- The containers and nutrient medium used to culture the microorganisms must be sterile.
- The food product is kept and packaged under sterile conditions and given a 'use-by date' before which the levels of microorganisms will be at a safe level. The product may have a final sterilisation to extend its shelf life.
- Microbiological testing takes place systematically at each stage of manufacture, to count the number of bacteria in a given amount of sample.

Working safely with microorganisms in school laboratories

Aseptic techniques are precautions that microbiologists take when handling microorganisms in order to protect themselves from harmful bacteria. The techniques also keep other bacteria out of the sample being investigated.

In school laboratories, bacteria are grown in Petri dishes on a culture medium solidified in agar gel. The culture medium provides them with the nutrients (energy source, minerals, supplementary protein and vitamins) needed for growth. It is possible to work safely with microorganisms in school labs, but you must observe sensible precautions when handling cultures. These are sometimes called **aseptic techniques**.

The most important precautions to take when working with bacteria are the following.

- Avoid contaminating a culture with 'wild' bacteria or human pathogens.
- Work at a temperature too low to grow pathogenic bacteria. In school labs, the incubation temperature should be kept at 25 °C. (Industrial labs often use higher temperatures to speed up the growth of their bacterial cultures, but they have very strict hygiene.)
- Avoid all hand-to-mouth operations such as biting fingernails, sucking pens or licking labels.
- Prevent direct contact with any bacteria being cultured. This means that Petri dish lids should be firmly sealed with tape and not opened.

Preventing contamination of your culture

An **autoclave** is an industrial steriliser. The high temperature kills microorganisms.

Here are some simple guidelines to prevent contamination taking place.

- Sterilise the Petri dishes and nutrient agar in an **autoclave**.
- Wash your hands and put on a clean lab coat to prevent contamination from your clothes.
- Close doors and windows to reduce air movement.
- Clear the bench where you are working and wipe it over with alcohol, which quickly destroys any bacteria.
- Lay out a clean paper hand-towel and collect all your equipment on this including the *closed* agar plates, culture, inoculating loop and forceps.

Making a streak plate

Microbiologists make streak plates to identify the types of bacteria present. An inoculating loop is used to spread the bacteria out over the surface of the agar. By the third or fourth streak, only a few bacteria cells are transferred. Circular colonies grow around single bacteria and this makes it easier to identify the different bacteria present.

The sequence of diagrams in Figure 11.28 shows how to set up a streak plate from a bacterial culture. This section also explains how to handle culture bottles to prevent contamination. Read the whole section before starting any practical work.

- Label the Petri dish on the underside with as little writing as possible so that you can easily see any growths.
- Light the Bunsen that you will use to sterilise the inoculating loop. Start heating at the handle end of the wire and move towards the loop end. Heat all parts of the wire until they are red hot (Figure 11.28a).
- Be patient: allow the loop to cool in the air below the flame before dipping it in the culture solution.

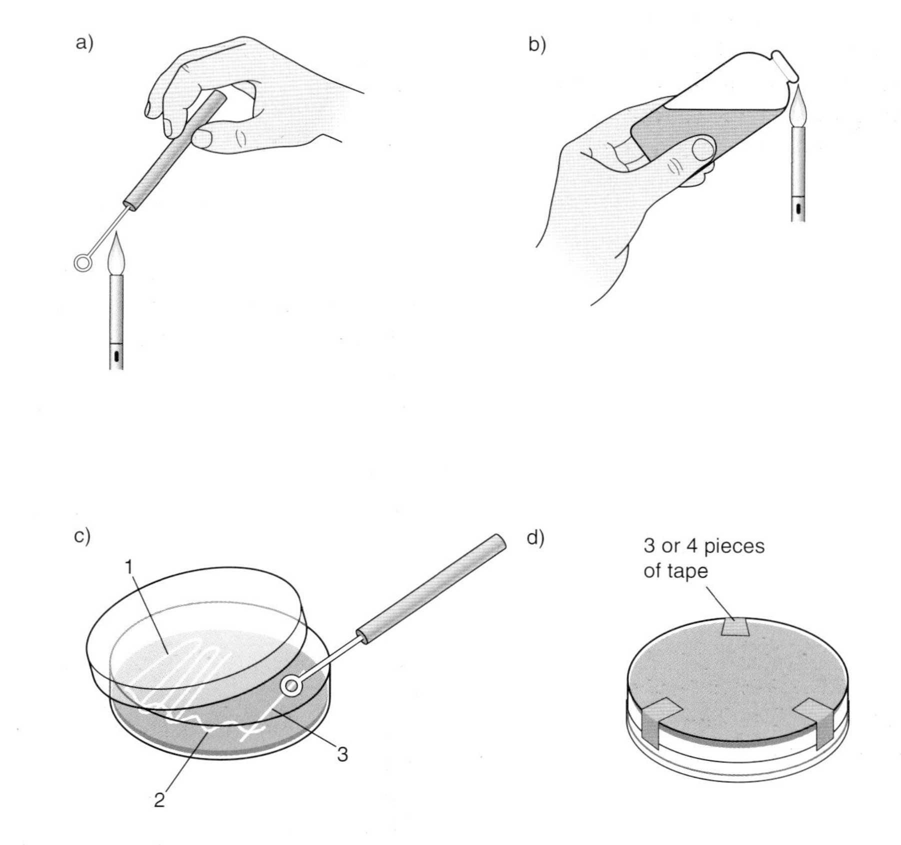

Figure 11.28 The procedure for culturing bacteria in the laboratory

SAFETY: Wear eye protection whe:. heating the wire loop.

A hot wire cannot transfer live bacteria and this is a common mistake made by students. If you touch the edge of the agar with the loop, and hear hissing, it is far too hot!

- Now, holding the loop with your thumb and first finger, open the lid of the culture bottle using your little finger. Draw the mouth of the open bottle through the flame (Figure 11.28b). Do not put the lid on the bench. Hold it between your little finger and the palm of your hand. Dip the loop in the culture, reflame the open bottle mouth and then replace the lid. In this way you should protect the culture bottle from contamination.
- Make streak 1 on one side of the agar plate without completely removing the lid of the plate. The streak should be made on the surface (not dug into the agar gel) as the bacteria are respiring aerobically (Figure 11.28c).
- Resterilise the loop, allow it to cool and then draw it through the tail end of the first streak. This will make a more dilute second streak.
- Repeat this dilution technique by resterilising the loop and drawing it through the tail end of the second streak to create a third streak.
- Place the loop in sterilising solution.
- Tape the lid on the plate and place it upside down in the warming cupboard to incubate until you are ready to observe the result (Figure 11.28d).
- Place all other equipment in sterilising solution. Roll up the paper towel for incineration. Wipe the bench with alcohol or disinfectant. Wash your hands and dry them on a paper towel.
- Before examining the plate, prepare your clean area as before. Do not open the plate; colonies should be visible through the base.
- When you have finished with the plate, place it in the disposal bag and it will then be autoclaved.

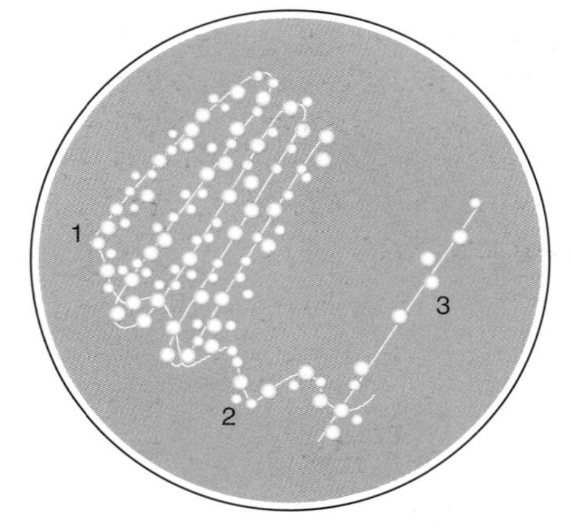

Figure 11.29 A streak plate seen from below showing thick growths of bacteria around streak 1, fewer bacteria around streak 2 and individual colonies around streak 3.

If you have made your streak plate successfully, the result should look like Figure 11.29.

23. In carrying out experiments with bacteria in the school laboratory, you should take precautions to avoid contamination by foreign bacteria. Give three possible sources of foreign bacteria.
24. When culturing bacteria in a school lab the temperature is kept at 25 °C, but when cultures are incubated in a hospital pathology lab the temperature is maintained at 37 °C. Explain the reasons for this difference in temperature.
25. What does agar gel contain?

Activity – Investigating the activity of different antibiotics

This practical was carried out by students who carefully followed all the aseptic guidelines for making a streak plate given in Section 11.8.

They collected the following items and placed them on a clean paper towel: Bunsen burner, fine forceps, and a labelling pen. They also had two Oxoid multodiscs or combi discs ready in clean empty Petri dishes. Each multodisc is a filter paper that has been impregnated with several different antibiotics and the small circles are numbered to identify the antibiotic present using a key (Figure 11.30).

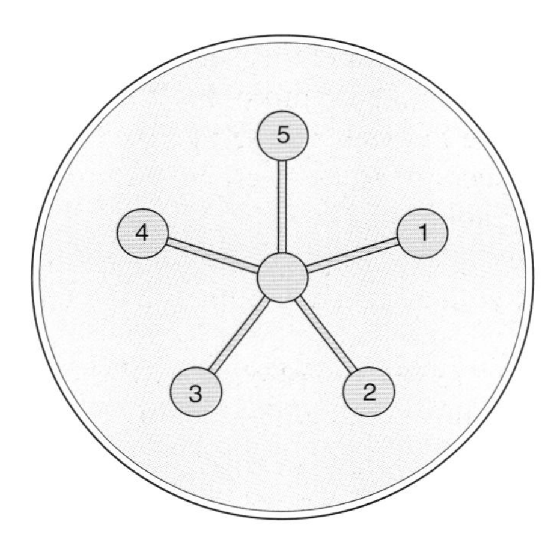

Figure 11.30 A multodisc

They were given two Petri dishes that had been seeded with bacteria all over the surface and labelled with the type of bacteria.

Procedure

- Students heat-sterilised the fine forceps in the Bunsen flame and allowed them to cool below the flame.
- They picked up the multodisc with the forceps and, lifting the lid of one of the Petri dishes, placed a multodisc on the surface of the bacterial culture. The disc was pressed into position using the forceps.
- The lid was sealed on the dish with strips of tape.
- They resterilised the forceps.
- This procedure was repeated with the second Petri dish.
- Both dishes were placed in the incubator until the next lesson.
- They followed aseptic techniques for clearing up.

1. How would you expect the antibiotic to spread from a multodisc disc into the agar?
2. When bacteria grow on agar in a Petri dish they appear as a white film. The different antibiotics affect different bacteria in different ways (Section 2.7). If an antibiotic prevents a bacterium from forming new cell walls during cell division, what would you expect to see around the discs?

Looking at the results

At the next lesson, the students set up a sterile area again and collected the Petri dishes, but did not remove the tape or open the dishes. They used a ruler or dividers to measure the diameter of the clear area around each disc and recorded their results for each bacterium and each antibiotic on Table 11.1.

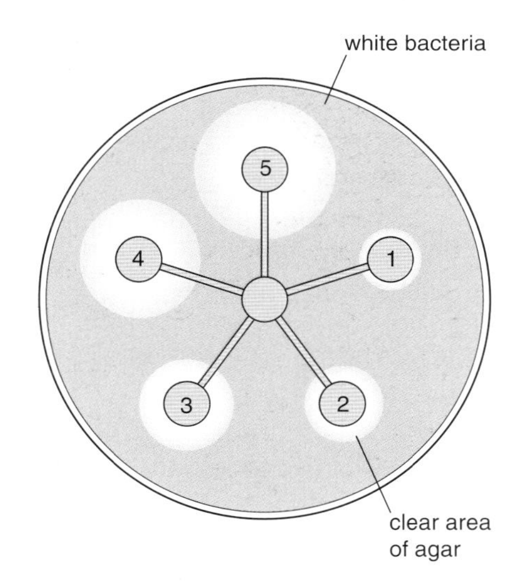

Figure 11.31 Example of the plate after incubation

When clearing up, they placed the used Petri dishes in the bag for autoclaving, cleared and disinfected the area and washed their hands.

Look at the results table. A bacterium is said to be susceptible to an antibiotic if the clear area has a diameter greater than 15 mm.

1. Were all antibiotics equally successful for each bacterium?
2. Give the number of the antibiotic that could be used to treat both types of bacterium.
3. Give the number of the antibiotic to which each of the bacteria is most sensitive.

	Diameter of clear area in mm				
	Disc 1	**Disc 2**	**Disc 3**	**Disc 4**	**Disc 5**
E. coli (K12)	12	23	17	15	9
Bacillus subtilus	28	13	22	7	11

Table 11.1 Results

Summary

- ✓ The development of the microscope enabled early scientists to observe microorganisms.
- ✓ Spallanzani, Schwann and Pasteur at different times did experiments that showed that the microorganisms that caused decay were not produced spontaneously.
- ✓ Pasteur showed that heating to high temperatures was sufficient to kill the microorganisms without spoiling the taste of a product. We now call this process **pasteurisation**.
- ✓ The idea of spontaneous generation was replaced by the **theory of biogenesis**.
- ✓ Modern cell theory provided an explanation of how living things are produced from other living things.
- ✓ Yeast is a single-celled organism that can respire both aerobically and anaerobically.
- ✓ Anaerobic respiration takes place without oxygen.
- ✓ Anaerobic respiration by yeast is called **fermentation** and produces carbon dioxide and ethanol.
- ✓ Sugars for fermentation of beer and wine are obtained from grain (beer) or grapes (wine). The starch in grain has to be first broken down by enzymes in a process called **malting**.
- ✓ In bread making, yeast enzymes convert starch to sugars and then break down the sugars anaerobically.
- ✓ The manufacture of yoghurt uses enzymes from bacteria to break down milk sugars (lactose) to lactic acid.
- ✓ Cheese manufacture also employs bacteria to produce lactic acid. In addition, enzymes break down the proteins and fats. Cheese making takes place at a lower temperature and is slower than making yoghurt.
- ✓ Pure cultures of microorganisms are important for food production.
- ✓ Contamination by other microorganisms that could be pathogenic is prevented by pasteurisation, sterilisation of equipment and sealing of containers.
- ✓ When working with bacteria in school laboratories, it is essential to use **aseptic techniques** to prevent contamination with pathogenic bacteria.
- ✓ Incubation temperatures in school should not exceed 25 °C to prevent pathogens from growing.
- ✓ Care must be taken to sterilise all equipment used. No culture plates should be opened and all plates should be **autoclaved** after use.

EXAMQUESTIONS

1 a) In the production of cheese, milk is heated to 73 °C for 20 seconds.
   i) What is the name of this process? *(1 mark)*
   ii) What is the purpose of this treatment? *(1 mark)*
b) The second stage of cheese production is the addition of a bacterial culture. This breaks down the milk sugar (lactose) to lactic acid.
   i) What is the effect of lactic acid on milk protein? *(1 mark)*
   ii) Rennet containing an active ingredient is also added at this stage. What is the function of this ingredient? *(1 mark)*
c) What are 'curds' and 'whey'? *(2 marks)*
d) What two factors help to preserve cheese during the slow ripening process? *(2 marks)*
e) The cheddar blocks are ripened at low temperatures. What enzyme reactions take place during this stage? *(2 marks)*
f) Supermarket cheddar cheese always tastes the same and has the same texture. How is this quality control achieved? *(2 marks)*

2 Look carefully at Figure 11.32.

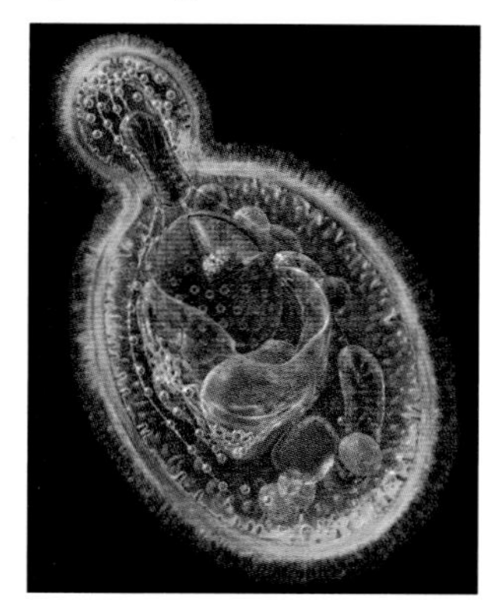

Figure 11.32 A photograph of a yeast cell taken with an electron microscope. The green outer part is the fungal cell wall, the purple area is the nucleus and the red object is a mitochondrion.

a) What type of respiration would you expect in a yeast cell containing an active mitochondrion? *(1 mark)*
b) What is happening at the top-left of the yeast cell? *(1 mark)*
c) Yeast cells were grown in a culture medium for 24 hours in a sealed flask at an optimum temperature. At two-hourly intervals a small sample was removed and the number of cells was counted. The results are shown in the table below.

Time in hours	Number of yeast cells in millions / cm^3
2	20
4	40
6	80
8	150
10	298
12	450
14	580
16	620
18	640
20	640
22	640
24	640

   i) Plot a graph of the number of yeast cells against the incubation time. *(2 marks)*
   ii) What do you notice about the increase in the number of cells for the first six hours? *(1 mark)*
   iii) What is happening to the division of the cells after 12 hours? *(1 mark)*
   iv) Suggest why the number of yeast cells becomes constant. *(1 mark)*
   v) What substance produced by the yeast inhibits reactions in the yeast cells? *(1 mark)*

3 a) What is the origin of the carbohydrate that yeast enzymes ferment to make beer? *(1 mark)*
b) How is this carbohydrate converted to a form that yeast can use? *(1 mark)*
c) In large-scale beer production there are two products in addition to the beer. Name these products and give a use for each. *(4 marks)*

4 Yoghurt is made by fermentation using bacteria.
a) What was the advantage of yoghurt to people living thousands of years ago? *(1 mark)*
b) Natural yoghurt has a sharp taste. What causes this sharp taste? *(1 mark)*

EXAMQUESTIONS

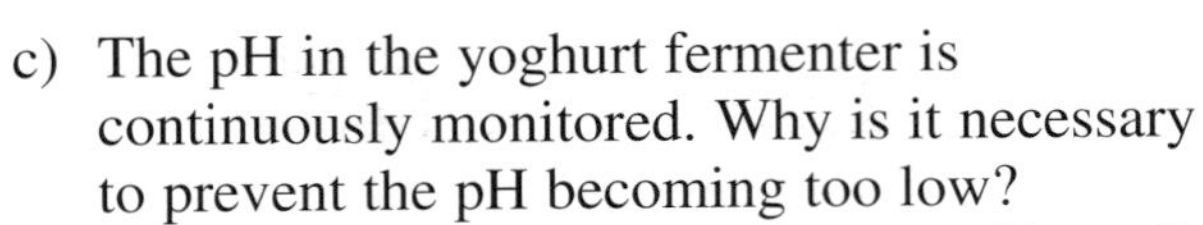

c) The pH in the yoghurt fermenter is continuously monitored. Why is it necessary to prevent the pH becoming too low? *(1 mark)*

d) At the end of the fermentation process, the temperature is reduced to 5 °C. What effect does this have on the enzyme reactions? *(1mark)*

e) One of the bacteria in yoghurt manufacture is *Streptococcus thermophilus*. 'Thermo' means heat; 'philus' means love. How does this relate to the optimum temperature in the incubation tank? *(1 mark)*

f) Why must the air bubbled through the incubation tank be sterile? *(1 mark)*

5. Figure 11.33 shows a Petri dish containing the bacterium *Bacillus subtilis* growing on nutrient agar jelly. The small white discs are pieces of filter paper each impregnated with a different antibiotic.

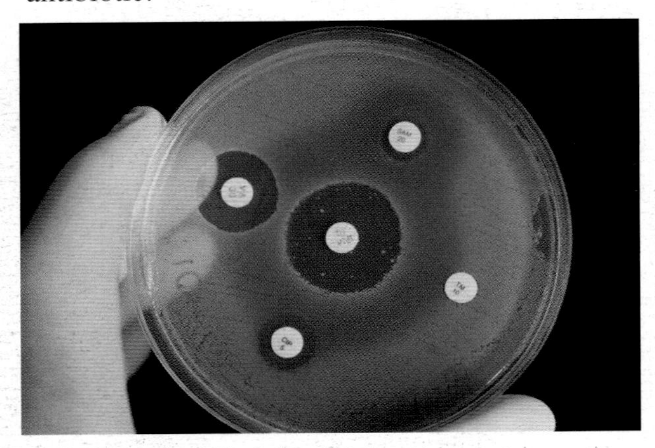

Figure 11.33 The effect of antibiotics on bacterial growth

The agar was seeded with the bacteria by pouring 1.0 cm^3 of *Bacillus subtilis* suspension onto its surface and allowing this to stand for 30 minutes. The antibiotic discs were then placed onto the surface using sterile forceps. The dish was taped and then incubated at 30 °C for 2 days.

a) Describe how the forceps were sterilised. *(1 mark)*

b) How does the antibiotic spread from the disc into the agar? *(1 mark)*

c) Explain:
   i) the white areas;
   ii) the clear areas around some of the antibiotic discs in Figure 11.33. *(2 marks)*

d) Why was the dish incubated? *(1 mark)*

e) Which antibiotic is *Bacillus subtilis* most sensitive to? *(1 mark)*

f) How could a doctor use this technique to prescribe a suitable antibiotic to cure a patient suffering from a sore throat? *(3 marks)*

Chapter 12
What other useful substances can we make using microorganisms?

At the end of this chapter you should:

- ✓ understand how biogas is produced by anaerobic fermentation of plant products or waste materials;
- ✓ be able to evaluate the advantages and disadvantages of both large-scale and small-scale biogas generators;
- ✓ appreciate that many different bacteria are involved in the breakdown of materials to produce biogas;
- ✓ know how ethanol is produced by anaerobic fermentation of carbohydrate materials;
- ✓ be able to discuss the economic and environmental advantages of using biofuels instead of fossil fuels;
- ✓ be able to describe the aerobic fermentation process for penicillin production;
- ✓ understand that conditions inside the fermenter must be carefully controlled for efficient production of penicillin;
- ✓ know how mycoprotein is produced using the fungus *Fusarium*;
- ✓ understand the differences between antibiotic and mycoprotein production;
- ✓ appreciate the need for pure cultures of microorganisms and sterile conditions when producing food or medicines.

Figure 12.1 Discovered nearly 80 years ago, the antibiotic penicillin is now produced industrially in large fermenters using the microorganism *Penicillium*. Notice all the equipment – sensor probes, inlet valves and so on – attached to the fermenter to control the process inside.

12.1 Producing biogas using microorganisms

Biogas is a mixture of gases, mainly methane and carbon dioxide (usually about 60% methane and 40% carbon dioxide). It is increasingly being used, particularly in developing countries, as a fuel for heating, cooking and generating electricity. Today, biogas provides about 1% of the UK fuel supply but about 45% of that in Sri Lanka.

1. Name the first gas produced in a biogas generator, just after the waste material has been added. Is this gas combustible?

Biogas is a mixture of gases produced when bacteria break down plant or animal waste. It is used as a heating fuel or to power electricity generators.

Anaerobic fermentation is fermentation without oxygen. Fermentation is a type of anaerobic respiration by microorganisms. The reaction produces gases and other products from the breakdown of carbohydrate.

How is biogas produced?

Biogas is produced when microorganisms decompose carbon-rich materials in the absence of air. In many rural areas, cattle dung is used as the raw material to produce biogas but any carbon-rich material is suitable. This could be human sewage, farm slurry (manure), restaurant scraps, chopped vegetation or any biodegradable materials. These waste materials are added to an enclosed tank known as a biogas generator or digester.

Biogas is produced by a process of **anaerobic fermentation**. Anaerobic means 'without air', so the process takes place in the absence of air. The microorganisms that break down the carbon-rich materials are bacteria that come from the soil or the intestines of animals and are found in animal dung. These bacteria function most efficiently at 30–40 °C and in anaerobic conditions. However, at the start of the process there is air in the tank, so the first gas produced is carbon dioxide. This is just how decomposer bacteria respire aerobically in an open compost heap. When most of the oxygen has been used up, a mixture of approximately 60% methane and 40% carbon dioxide is produced. Because the bacteria produce methane as well as carbon dioxide, they are described as methanogenic.

Biogas generators

The original biogas generators consisted of a single clay-lined pit with a narrow inlet for the raw material (Figure 12.3). The gas produced was collected from the top of the pit and piped to the user. Notice in Figure 12.3 that the bend in the inlet pipe prevents air entering the pit, if the level of liquid is high enough. This is important to maintain anaerobic conditions. The main problems were leakage of gas if the top of the clay lining dried and cracked, and a gas supply that was not continuous. When the production rate dropped, the tank had to be emptied and cleaned out.

Figure 12.2 A small-scale biogas generator in India. The manure and vegetable waste are mixed, then added to the digester through the brick-built chamber (right). The dome of the gas collector expands as biogas is produced.

Modern biogas generators (Figure 12.4) are larger and have three tanks. Animal manure and chopped vegetation are added to a collecting and mixing tank. From here, an inlet pipe takes the material to the bottom of the slurry in the generator, which is sealed off from the air. The generator tank is constructed from bricks and mortar or from polyethylene. As the biogas is produced, it collects in the expandable dome of the generator tank above the liquid slurry. When biogas is being produced, the pressure of the biogas forces slurry down the outlet pipe into a third tank, from which it can be removed and used as liquid manure on the fields. Unlike early biogas generators, this results in a

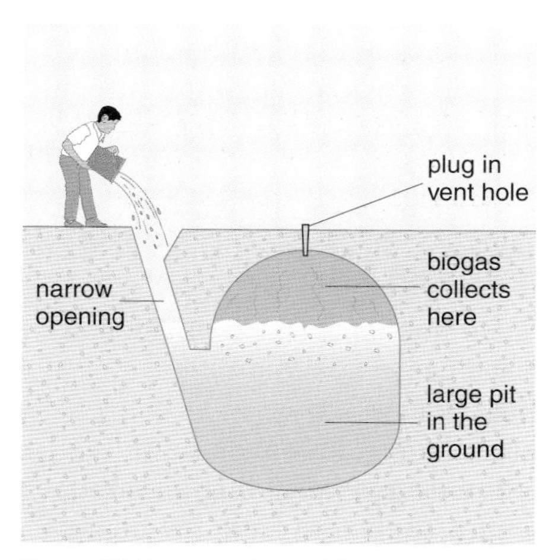

Figure 12.3 An early rural biogas generator

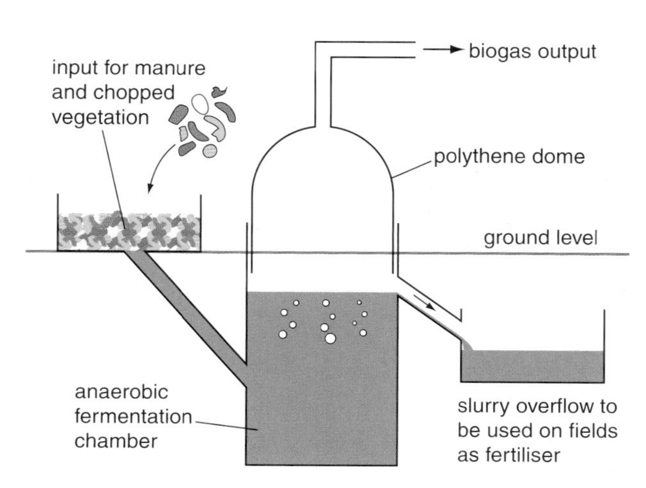

Figure 12.4 The design of a modern biogas generator produces a better yield of biogas, and needs less maintenance.

semi-continuous process rather than a batch process. Using a modern generator, the manure from two dairy cows can provide a fairly continuous supply of biogas to meet the cooking and lighting needs of a small family.

Types of bacteria in biogas production

Bacteria in a biogas generator first break down the raw materials to products on which the methanogenic bacteria can get to work. The species of bacteria needed to start the breakdown depends upon the raw materials added to the generator. In a biogas generator supplied with farm waste such as chopped straw and cow dung, bacteria in the first stage of the process must produce enzymes that break down cellulose to sugars and finally glucose. In the next stage, other bacteria produce enzymes that convert the sugars to other carbon compounds. The final group, methanogenic bacteria, produce enzymes to convert these compounds to methane. If there are proteins in the farm waste these are broken down by bacteria that produce protein-digesting enzymes.

Each type of bacterium in a biogas generator functions under its own optimum conditions. For example, methanogenic bacteria cannot function at very low pH. However, the drop in pH caused by organic acids is neutralised by ammonia from protein breakdown, so the methanogenic bacteria can survive. Gas production is also dependent upon temperature. The bacteria function best at about 35 °C, so biogas is not as easy to make in northern Europe as in hot countries. Fortunately, the fermentation reaction in the generator is exothermic, and the generator can be insulated to keep this heat in.

Environmental, social and health advantages of using biogas

Before biogas, the main fuels used in developing countries for cooking and heating were wood and dried cattle dung. These fuels have several disadvantages, compared with biogas:

Figure 12.5 A woman in India cooks at a stove powered by biogas.

- The wood for fuel is taken from forests, causing trees to be damaged or lost. Over time, loss of trees causes soil erosion, making it difficult to grow other plants. Biogas uses waste materials rather than wood and so allows trees and scrubland to regenerate.
- To collect wood for fuel, people need to make long, exhausting journeys carrying heavy bundles of wood. Biogas uses materials that are closer to hand.
- Burning wood inside a poorly ventilated home produces smoke and fine particles that can damage lungs and cause health problems. Cooking on biogas stoves is clean and odourless.
- Using dried cattle dung as a fuel means there is less manure to improve the soil, and so crop production decreases. While biogas can also be produced using dung and manure, the waste from the biogas generator can be used as fertiliser, increasing crop productivity.
- Where a rainwater collection system is modern, flushing toilets can be installed. This produces better sanitation and consequently a general improvement in health.

A **carbon-neutral** fuel causes no overall increase in carbon dioxide levels in the atmosphere when it is burned, because the carbon dioxide taken up during photosynthesis equals that produced by combustion.

Biogas, like wood, is formed from a renewable source – it will never run out. Biogas is also a **carbon-neutral** fuel that does not contribute to increased carbon dioxide emissions. This is because, even though carbon dioxide is produced in the decomposition process and again when biogas is burned, carbon dioxide was removed from the air when the original plant material was growing.

A web search for Ashden Awards will provide more information on this and other 2006 projects and interesting comments from users about the advantages.

2. State one economic advantage of using biogas.
3. State four ways in which the health and lifestyle of people in developing countries improved as a result of the use of biogas.
4. What are the benefits to the environment of using biogas as a fuel in countries such as Nepal?

Activity – Biogas comes to Indian towns

Biogas generators are now common in rural areas of India where there are cattle providing lots of manure. Unfortunately the conventional type of generator cannot be used in towns and cities, where there are no livestock. However, Dr Karve from the Appropriate Rural Technology Institute (ARTI) in India invented a small generator that could be used in urban areas. In 2006, he won the Ashden Award for renewable energy.

A conventional biogas generator, sunk into the ground, works on human or animal excreta and starts producing sufficient gas after 40 days. Dr Karve's new compact generator can use any starchy, cellulose-rich or sugary vegetable food waste, and gas production starts after only 2 days. What is more, this generator produces 1.0 m^3 of biogas from only 2 kg of waste, compared with 80 kg of manure needed to produce 1.0 m^3 of biogas in a traditional generator; 2 kg of waste is easily achieved with fruit and vegetable skins,

Figure 12.6 The proud owner standing next to his urban biogas generator. The generator is small enough to fit in an urban garden or terrace.

spoilt milk and food leftovers and this produces enough gas for cooking each day by the average urban family. The gas produced is carried from the generator, situated just outside the house, directly into the kitchen biogas stove. A biogas generator replaces the need for cylinders of 'bottled' gas, which are expensive, heavy and need collecting or delivering.

1. How does the structure of the new biogas generator compare with the conventional rural type?
2. What are the advantages of the new compact biogas generator for urban populations?
3. Why can the urban biogas generator be smaller?
4. In addition to the production of biogas for cooking, what is the main environmental advantage of the new generator?
5. a) State the typical composition of biogas.
   b) Assuming 60 : 40 composition, what volume of methane will be present in 500 dm^3 of biogas?
   c) Why do you think the energy produced from biogas can vary?
   d) Why do you think a typical cooker has an efficiency of only about 60% when using biogas compared with methane?
6. The fuel that biogas replaces in urban areas is 'bottled gas', or methane. The energy value of methane is 36 120 kJ / m^3.
   The energy values of biogas are typically:

   21 500 kJ / m^3 for 60% methane;

   25 200 kJ / m^3 for 70% methane.

   a) Using these figures, complete Table 12.1 to calculate the volume of gas required to boil 1 dm^3 of water using biogas with different methane contents. Copy the table and then enter the values. Assume the starting temperature of the water is 20 °C and that 4.2 kJ raises the temperature of 1 dm^3 of water by 1 °C.
   b) Why will a family benefit from including waste with a higher carbon content?
7. A typical biogas cooker uses 0.3 m^3 per hour and requires 2 kg of feedstock to produce 1.0 m^3 of biogas.
   What mass of feedstock is required to provide sufficient gas for 3 hours of cooking?
8. Suggest two factors that will affect the efficiency of a biogas stove.

	Biogas with 60% methane	Biogas with 70% methane
Temperature through which the water has to be raised in °C		
Energy required to raise 1 dm^3 water to its boiling point in kJ		
Volume of gas required = (energy required in kJ) / (energy value of the fuel)		

Table 12.1 A comparison of biogases

Using large-scale biogas generators for waste disposal

Figure 12.7 This biogas plant produces enough biogas to generate electricity for a village in the Netherlands. The circular slurry tank contains raw manure that is fed into the large anaerobic digester.

Large biogas generators can solve the waste-disposal problems of intensive pig and poultry farms. Pigs produce a large volume of waste, which must not be allowed to pollute water. The problem is severe in countries such as Holland and Denmark where there is intensive pig production but not enough farmland on which to spray all the manure produced. Other problems can result if heavy rainfall washes the manure into rivers, where it causes environmental damage through a process called eutrophication (Section 3.3). In this process, extra nitrates in the manure cause rapid growth of algae and the water becomes green. As the algae die, aerobic bacteria decompose them and the water becomes brown, anaerobic and very smelly.

The large biogas generator in Figure 12.7 operates almost entirely on pig slurry. This is collected in tanks before being fed into the generator. The generator is kept at an optimum temperature by burning some of the biogas produced. The biogas can also be burned to dry the final product to make a dry fertiliser. Gas production is large enough to heat the farm buildings and power an electricity generator.

In addition to using up excess manure, large-scale biogas generators can reduce the need for landfill (Section 3.3). Because of this, the number of industrial-scale biogas generators that can use domestic refuse is increasing. The basic process is the same as in small-scale biogas generators. Mixed biodegradable rubbish is fed through a machine which chops it into small pieces before adding it to the anaerobic digester. The biogas produced can be used for local heating. Liquid waste from the digester can be used on the land and any solid material can be pasteurised and used as a solid fertiliser for gardening.

5. Why must the final product from a refuse biodigester be pasteurised before it is used on greenhouse crops?
6. If a biogas digester uses restaurant waste as a raw material, suggest what enzymes would need to be produced by the bacteria for digestion of the material.
7. What are the environmental and economic benefits of using a biogas generator for intensive pig farms?
8. How would the conditions in a biogas generator in Scotland be kept at optimum levels in the winter?

Producing bioethanol using microorganisms

Bioethanol (spirit fuel) is ethanol produced by the anaerobic fermentation of carbon-based material. The term is sometimes shortened to 'biohol'.

Apart from biogas, there are other important fuels produced by microorganisms. These are based on ethanol and are usually called spirit fuel or **bioethanol**.

Ethanol-based fuels are produced by anaerobic fermentation of sugars. The sugars can be obtained directly from plants such as sugar cane and sugar beet, or made by conversion of starchy plant materials such as maize (sweetcorn). Currently, normal petrol vehicles can run on petrol with 5% ethanol, although specially modified models can use fuel that is 85% ethanol and only 15% petrol. The first bioethanol pump supplying car fuel in the UK was opened in March 2006.

Figure 12.8 This modified Saab 9-5 is designed to run on 15% petrol and 85% bioethanol. The higher octane of this fuel gives the car a 20% increase in brake horsepower – with lower carbon dioxide emissions

Brazil is the leading bioethanol producer at present. The country started developing alcohol fuels as an alternative to petrol after the 1975 oil crisis, when petrol and other oil products were in very short supply. Brazil now exports a significant amount of bioethanol to other countries. The climate in the south of Brazil is suited to the production of sugar cane and half of the yield is used for bioethanol production. There are no petrol-only cars in Brazil. All of them have engines designed to run on a blend of ethanol with petrol.

How is bioethanol made from starchy crops?

There are four stages in the production of bioethanol from maize or grain (wheat, barley or rye).

- The first stage is milling (crushing). Before milling, the grain or maize is soaked in water to soften it. The milled grain is then separated into three products: the 'germ', which is used to make corn oil; the fibre and the starch. Only the starch is used in bioethanol production.
- In the second stage, carbohydrase enzymes (Section 7.5) are used to convert the starch to sugar.
- The third stage is fermentation of the sugars produced. Yeast is added to the sugar solution to provide the enzyme invertase, which converts sucrose in the sugar solution to glucose and fructose. Another enzyme from the yeast called zymase converts the glucose and fructose to ethanol.

$$\text{sucrose} + \text{water} \xrightarrow{\text{enzyme invertase}} \text{glucose} + \text{fructose}$$

$$\text{glucose or fructose} \xrightarrow{\text{enzyme zymase}} \text{ethanol}$$

- In the final stage, ethanol is separated from the solution by **fractional distillation**. The alcohol is driven off at 78 °C and condensed.

Fractional distillation is the separation of a mixture of liquids by evaporating the liquid and then condensing the vapour. The liquid with the lowest boiling point condenses first.

9 Name the three enzymes and write word equations for the reactions involved in the production of ethanol from barley grains.

Quote: 'One hectare of wheat produces about 29 000 miles of motoring, enough to take a car around the equator and still have 4000 miles of fuel left.' *Green Spirit Fuels*

What are the advantages of bioethanol?

Bioethanol has quite a few impressive advantages, both in production and in use. Look at the list below:

- Bioethanol is considered to be carbon neutral because the carbon dioxide produced when it is burned is balanced by the carbon dioxide used in photosynthesis by the plants. However, we must add to this balance any carbon dioxide produced from the fuel used in the industrial production processes such as milling and distillation. Even so, bioethanol made from grain produces 65% fewer greenhouse gases than petrol, according to UK Government figures.
- Using bioethanol reduces pollution from car exhaust fumes because it promotes complete combustion of the fuel within the engine forming carbon dioxide, and no poisonous carbon monoxide or unburnt hydrocarbons.
- If bioethanol is spilt in a tanker accident, the ethanol can be diluted with water to non-toxic levels and is biodegradable. This is not true for petrol or diesel spills, which can cause severe environmental damage.
- Bioethanol is formed from renewable sources and this conserves the supply of fossil fuels (oil, gas and coal).
- Using bioethanol reduces a country's dependence on fossil fuel imports. In the UK, crops such as sugar beet and grain can be grown and used for fuel production, reducing the need to import fuel from other countries. In the future, improved technology may allow the production of ethanol from quick-growing woody plants.
- The production of crops and the bioethanol-manufacturing processes have already created many new jobs in Brazil, Germany and regions of the UK such as East Anglia.
- Bioethanol is easy to use as a car fuel as it can be distributed, stored and delivered from forecourt pumps in the same way as petrol.

10 Explain what is meant by a 'carbon-neutral' fuel.

So, could bioethanol provide a solution to our energy problems? 'Going green' now by using carbon-neutral bio-fuels would reduce carbon dioxide emissions and help to reduce the greenhouse effect. But, in the longer term, bioethanol will not completely solve either the fossil fuel supply problem or the greenhouse gas problems, because worldwide there is not enough free land to grow the necessary crops.

One concern is that production of bio-fuel crops will reduce the food supply for local people in poorer countries, as the land is used to grow crops for fuel rather than for food. But, if a properly managed scheme were introduced, it would produce more employment and therefore better incomes and more food for local people.

Increased bioethanol production could also have some serious negative effects on the environment, unless it is managed properly. Allowing people to destroy natural habitats for the production of bio-fuel crops would not be beneficial. Instead, the planting of bio-fuel crops should be integrated into the management of existing agricultural land, minimising habitat destruction. Well-planned schemes can have a real benefit to the environment. For example, adding sugar cane production to a crop rotation scheme that includes a leguminous crop will keep the soil fertile. The yeasty waste products from sugar cane fermentation can be used as a nutrient-enriched food for dairy cattle. In turn, manure from the cattle can be used for soil enrichment.

Activity – What are the economic aspects of using bio-fuel in the UK?

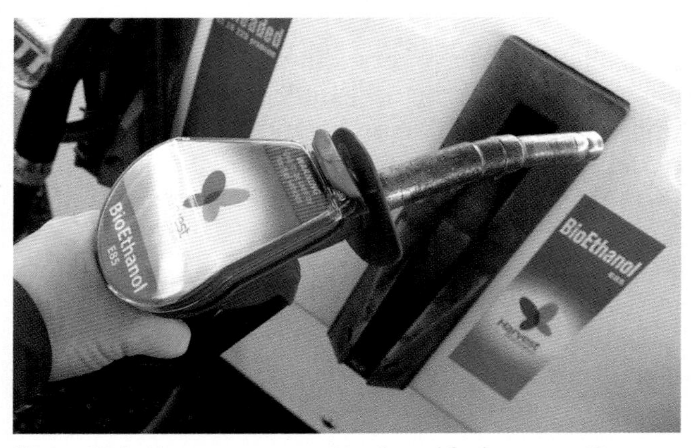

Figure 12.9 Have you seen a bioethanol fuel pump yet?

There are both benefits and drawbacks to the increased use of bio-fuels. Below are some points to discuss in a small group. After your discussions, produce a poster or presentation with the title 'The future of bioethanol in the UK'.

- Currently, most petrol stations are operated by firms like BP (British Petroleum). What difference might the sale of bioethanol make to their profits?
- How much are you influenced by 'green' policies?
- Bioethanol is expensive to produce but the price would be fairly stable, in contrast to the price of crude oil during world crises. Do you think that stable fuel costs would influence the development and sale of bioethanol?
- There is usually a grain surplus in the UK and in other European countries, which could be used for bioethanol instead of export. Could this reduce our dependence on oil imports?
- 'Set-aside' land in the UK (land not being used for food crops) could be used to produce fuel crops. What might be the possible economic benefits of this to people living in rural areas?
- In Sweden, there has been a big switch to bioethanol / petrol mixtures. Which of the following factors do you think might have caused this: i) environmentally aware consumers, ii) financial subsidies to bioethanol producers or iii) EU targets to reduce CO_2 emissions?

⓫ Bioethanol has been suggested as part of the solution to the fossil fuel crisis. What other developments will help to solve this crisis?

⓬ Suggest four advantages of the use of bioethanol in urban areas.

Using a microorganism to produce penicillin

You have probably heard of penicillin, a widely used antibiotic. The development of this hugely important medicine began in 1928, when Scottish bacteriologist Alexander Fleming made a discovery that would revolutionise the way doctors could treat infections. Fleming observed that a mould had developed on a discarded culture plate, and that the mould had resulted in a clear area around it. Something had killed the *Staphylococcus* bacteria growing close to the mould. Fleming carried out further experiments with solutions made from this mould, *Penicillium notatum*. Even when he used very dilute solutions, Fleming found these prevented the growth of *Staphylococcus* bacteria. He named the ingredient that was having this effect 'penicillin'.

Figure 12.10 A colony of *Penicillium* mould is growing on the nutrient medium in this Petri dish. Notice that there are no bacteria growing around the mould.

Fleming was unable to make further progress in developing penicillin as a medicine because he did not have the necessary techniques to extract and purify the key ingredient. Eleven years later, Howard Florey and Ernst Chain were trying to find new medications to treat soldiers injured during the Second World War. They had read Fleming's reports of penicillin's action as a **bactericide** so they isolated the penicillin in a pure form. Using this purified extract, they were able to carry out the first trials on patients. The trials proved sucessful: penicillin injections could cure wound infections that previously had caused death. However, the problem of producing penicillin in sufficient quantities remained for some years.

A **bactericide** is a substance that kills bacteria.

Producing penicillin on the large scale

Figure 12.11 Factory production of the antibiotic penicillin. The worker inspecting the fermentation tanks gives a good indication of their size.

Strictly speaking, fermentation is an anaerobic process carried out in the absence of air or oxygen. However, when living microorganisms are used to produce a product (such as an antibiotic), the process is often referred to as fermentation – even if the process is aerobic and requires oxygen.

Initially, penicillin was produced in flat culture dishes but this supply could not satisfy demand. In 1942, an existing deep-culture fermentation plant was modified to produce penicillin and the changes resulted in a major technological breakthrough. Within a year, millions of doses were being produced. In 1945, Alexander Fleming, Ernst Chain and Howard Florey jointly received a Nobel prize for the discovery of penicillin and its use in treating bacterial infections.

Large-scale production of penicillin requires sterile conditions, containers called fermenters (Figures 12.1 and 12.11) that hold a large volume of nutrient medium, and aeration with sterile air as the process is aerobic. Today, *Penicillium* strains that have been genetically modified to produce high yields are used. This is a batch process with several stages:

- A sterile sample of a mutant *Penicillium* strain is produced, and when this is growing well it is added to the culture medium. The sterile growth medium contains the substances necessary for rapid growth of the mould: a sugar such as lactose, amino acids, mineral ions and certain vitamins. The pH and temperature are set at the best levels to achieve the maximum growth rate of the mould.
- When the mould is sufficiently concentrated and the nutrients have been used up, the temperature is changed for the next stage of fermentation. In this later stage, when the *Penicillium* has used up most of the nutrients for growth, it starts to produce penicillin. Penicillin is formed as a secondary product, from the breakdown of a compound already produced in the cytoplasm of the mould.
- After about a week from the start of the process, the maximum antibiotic concentration is reached and the mould is filtered from the fermentation liquid. The antibiotic is separated from the liquid and then purified and crystallised.
- Finally, the antibiotic is processed into the form in which it is to be sold and used.

All these processes must take place under sterile conditions. In addition, penicillin is easily damaged by heat so the temperature has to be controlled throughout.

13. What prevented Fleming from developing the antibiotic after he had identified the action of penicillin as a bactericide?

14. Suggest possible consequences if the penicillin fermenter is contaminated by an unwanted microorganism. (You may need to revisit Chapter 11.)

Activity – Maintaining optimum conditions in a penicillin fermenter

Having read about the production of penicillin, look carefully at Figures 12.11 and 12.12. Then answer the following questions.

1. The temperature for fermentation is controlled by pumping cold water through the water jacket surrounding the fermenter. Why do you think it is necessary to pump cold water through the water jacket?
2. a) What is the purpose of the sensor probes that link to the computer in the diagram?
   b) What does the computer control?
3. The nutrient medium contains lactose, amino acids and mineral ions. Suggest how each nutrient is used by the mould.

4. The incoming air passes through a biological filter. What do you think this removes?
5. a) Why is air passed into the fermenter?
   b) Why is the concentration of dissolved oxygen monitored?
6. What are the purposes of the bubble plate and the stirrer paddles?

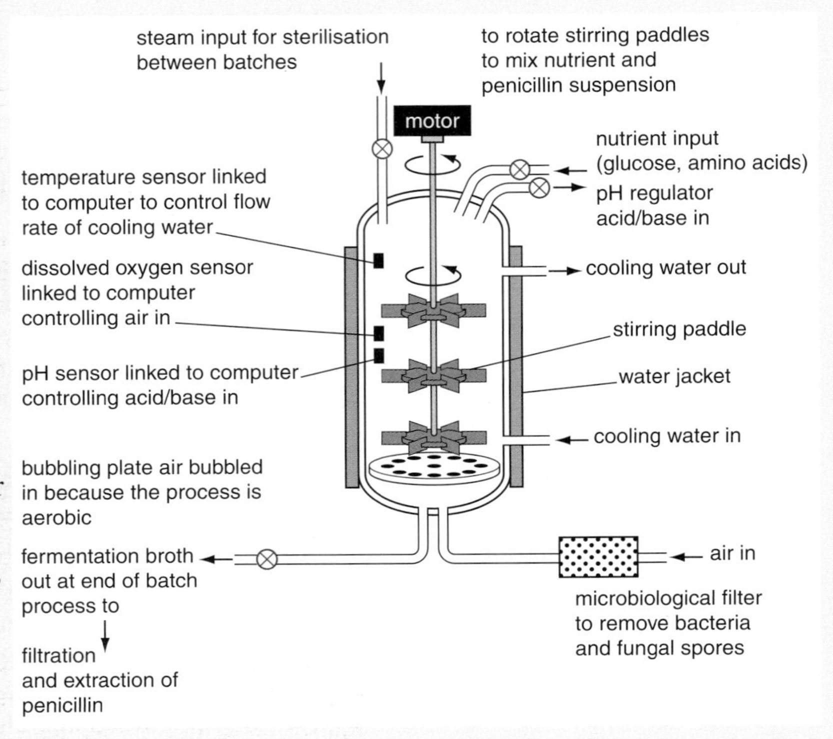

Figure 12.12 The control of the fermentation process to produce penicillin

12.4 Fungus food: producing Quorn by aerobic fermentation

Figure 12.13 Quorn is an edible substance made from mycoprotein. It can be processed to form a variety of different foods.

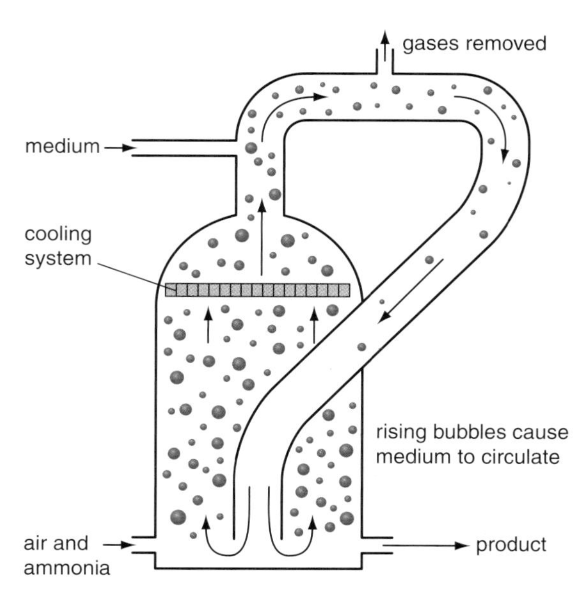

Figure 12.14 A fermenter for the production of mycoprotein

If you are a vegetarian, you will probably have heard of a food called Quorn™. Quorn is a protein-rich food that is produced using microorganisms. It has the advantage of being low fat and low calorie so many non-vegetarians and slimmers eat it as part of their diet. The advantage of Quorn in cooking is that its texture resembles meat and it can be substituted for meat in many recipes.

Mycoprotein is formed from the mycelium of a fungus. A mycelium is the mass of fine fungal threads called 'hyphae'.

Quorn is made from **mycoprotein** produced from the fungus *Fusarium,* which is found in the soil. The fermenter used in the process is a tall stainless steel container (about 40 m in height). The inside is kept sterile. A computer controls conditions so that they are optimal for the growth of the *Fusarium,* with the temperature maintained at 32 °C and the pH at 6. The sterilised nutrients are glucose (formed by breaking down starch using enzymes), mineral salts and vitamins. Sterilised air with some ammonia is pumped into the fermenter. The ammonia is a source of nitrogen for making proteins with the glucose. The conversion of glucose to protein is fast, and it takes only five hours for the biomass to double. A continuous feed of nutrients and simultaneous removal of broth containing the fungal threads results in a continuous fermentation process. The fungal threads are then collected by centrifugation and dried. After this, shaping and flavouring can take place.

There are two main differences between mycoprotein production and that of antibiotics such as penicillin:

- Mycoprotein is produced by a continuous-flow process, whereas penicillin is produced by a batch process.
- The fungal biomass grown is the required product for mycoprotein, whereas penicillin is obtained as a secondary product after the main fermentation process.

15. Mineral salts added to the mycoprotein fermenter include potassium, magnesium, phosphates and trace elements. What is the value to the consumer of adding these:
    a) to the *Fusarium*;
    b) to the Quorn?
16. What use do you think the *Fusarium* makes from the ammonia pumped into the fermenter?
17. The process for the production of mycoprotein is described as a continuous-flow process, compared with the batch process for penicillin. Explain what this means.

Using microorganisms safely

In the production of both antibiotics and mycoprotein, it is essential that pure cultures of the microorganism are used and that no unwanted microorganisms enter the fermenter. So, the tanks go through cycles of steam sterilisation after each production cycle. All nutrients must be sterilised and the air bubbled in must be micro-filtered to remove bacteria or fungal spores.

18. After a production run of about six weeks the mycoprotein fermenter is steam sterilised. Suggest why this is necessary.
19. Mycoprotein was initially developed as a food because of fears about protein shortages in poor countries. Suggest why this technology has not yet passed to poor countries.

Summary

- ✓ Anaerobic processes take place without oxygen, in the absence of air.
- ✓ Fermentation takes place using enzymes produced by bacteria or yeasts.
- ✓ **Biogas** and **bioethanol** are fuels produced by anaerobic processes.
- ✓ The typical composition of biogas is 60% methane and 40% carbon dioxide, but this varies according to the material fermented.
- ✓ Biogas generators break down organic waste material, producing biogas which is used for fuel.
- ✓ Many different microorganisms are involved in the breakdown of materials in biogas production.
- ✓ Biogas from small-scale or farm-size generators can meet energy needs, improve people's health and hygiene, increase crop productivity and give people more free time to earn a living.
- ✓ The technology can also be used on a large scale to break down household waste, reduce the need for landfill or incineration and convert the waste into fuel.
- ✓ Bio-fuels are **carbon neutral**. They remove the same mass of carbon dioxide from the atmosphere during photosynthesis as they return to the air when burned.
- ✓ **Bioethanol** is produced by the anaerobic fermentation of sugar obtained from plants. The main sources of sugar are sugar cane and sugar beet. Production of bioethanol from maize or grain requires the breakdown of cellulose to sugars using the enzyme carbohydrase as the first step. In the final step, ethanol is separated from the fermented solution by **fractional distillation**.
- ✓ Road vehicles can be modified to run on a mixture of petrol and ethanol.
- ✓ Bioethanol production reduces dependence on fossil fuels and therefore has both economic and environmental advantages.
- ✓ Antibiotics like penicillin are chemicals produced by moulds that kill or inhibit the growth of some bacteria. A **bactericide** is a substance that kills bacteria.
- ✓ Antibiotics are produced by aerobic processes in huge sterile fermenters. The fermenters are computer controlled to maintain optimum conditions of temperature, pH, aeration and nutrients such as sugar and nitrogen compounds. Aeration provides oxygen for respiration of the microorganisms.
- ✓ In industrial penicillin fermenters, the nutrient supply is deliberately limited to force *Penicillium* to produce penicillin.
- ✓ **Mycoprotein** is produced as a food from the mycelium of the fungus *Fusarium*. The conditions in the fermenter are computer controlled.
- ✓ Antibiotics are generally batch produced, whereas mycoprotein production is a continuous-flow process.
- ✓ Maintaining sterile conditions and pure cultures are essential for both antibiotic and food production, and when using microorganisms in the laboratory.

EXAMQUESTIONS

1 Biogas production takes place in the UK on farms and in municipal waste composting.

a) What is the typical composition of biogas? *(1 mark)*

b) Why is the heat produced by the combustion of biogas less than that from the combustion of methane? *(1 mark)*

c) State two advantages of the use of biogas generators on farms. *(2 marks)*

d) State two advantages of biowaste digesters in urban situations. *(2 marks)*

e) Explain the effect on gas production when the temperature in a biogas generator falls. *(2 marks)*

2 Bioethanol is produced by the anaerobic fermentation of carbohydrate crops such as grain.

a) Copy and complete Table 12.2 to explain the process. *(8 marks)*

b) What is meant by the term 'anaerobic'? *(1 mark)*

c) Where do the enzymes invertase and zymase come from? *(1 mark)*

d) What are the products of the complete combustion of bioethanol? *(2 marks)*

e) Why is bioethanol combustion safer than that of petrol? *(2 marks)*

3 Read the following text and answer the questions below about penicillin production.

The production of antibiotics has several stages. Following the identification of a species of mould that produces the 'active' chemical, a pure culture has to be grown under optimum conditions with suitable nutrients. This produces spores, which can be kept in an inert state until required. The spores are then added to a suitable culture medium to produce sufficient mould biomass to add to the fermenter vessel. The production vessel provides the conditions and nutrients for rapid growth. Once maximum growth has been achieved, changes in temperature and nutrient composition cause 'stress' to the mould. As a result of these changes, the mould further metabolises some of the substances that it has already made and stored in the cytoplasm. The product of these changes is the required antibiotic. Once the antibiotic is sufficiently concentrated, it can be harvested. The antibiotic is separated from the cooled filtrate using a solvent, then dissolved into water and collected by precipitation and filtration. Finally the antibiotic is vacuum dried.

a) Why is it essential to obtain a pure culture? *(2 marks)*

b) What conditions are maintained at optimum levels for maximum growth of the mould? *(3 marks)*

c) The nutrient medium contains: i) a carbon source; ii) a nitrogen source; iii) phosphates; iv) trace elements. Suggest why each of these is needed. *(4 marks)*

d) The antibiotic is a secondary product. What does this mean? *(2 marks)*

e) What is the first stage in harvesting? *(1 mark)*

Process	What happens in the process?	Result
Breakdown of the grain: i) Soaking ii) Milling iii) Hydrolysis	Grains are soaked. Milling breaks up the grains. The starch is __________ using carbohydrase enzymes	The three parts of the grain – the '______', the ________ and the ______ – are separated. __________ are formed.
Fermentation	Yeast is added containing the enzyme invertase and the enzyme zymase.	Converts sucrose to ____________. Converts _________ to ___________.
Distillation	Solution heated to ______ °C.	Ethanol driven off and then ____________________.

Table 12.2

Index